NUMBER

1 2 3 4 5 6 7 8 9 10

NUMBER

John McLeish

BLOOMSBURY

Author's Note

My son Kenneth McLeish edited the original text of this book, and prepared it for press. My grandson Simon McLeish edited the mathematical tables and workings, and researched the illustrations and bibliography. I am warmly grateful for their assistance. All opinions, judgements and choices remain my own.

First published 1991 by Bloomsbury Publishing Limited, 2 Soho Square, London W1V 5DE

A CIP record for this book is available from the British Library

ISBN 0 7475 0921 2

10 9 8 7 6 5 4 3 2 1

Designed by Geoff Green
Typeset by Vision Typesetting, Manchester
Printed and bound in Great Britain by
Butler & Tanner Ltd, Frome and London

CONTENTS

If a man's wit be wandering, let him study the
Mathematics

Francis Bacon Essays (1625)

INTRODUCTION

What song the Sirens sang, or what name Achilles assumed when he hid himself among women, though puzzling questions, are not beyond all conjecture.   *Sir Thomas Browne*

Number: the language of the exact sciences

The binary assumption, which in computer science says that switches can only be either ON or OFF, suggests that people might also be divided into two kinds: those who are 'turned on' or excited by numbers and those who are depressed by them. The second group is the vast majority. Their response was once summed up by an adult student, university-educated, who told me after I had tried to demonstrate something on the blackboard with numbers all less than nine, 'As soon as you write a number on the blackboard – any number – I have such a feeling of sick inadequacy that I can't think any more'.

It must be said from the outset that there is nothing inherently sadistic about either numbers or those who habitually use them or teach about them. Neither is it true – binary assumption notwithstanding – that you either have a natural gift for mathematics, or you do not. The only real distinction to be made here is between those who were taught number badly and those whose teachers understood that mathematical ability is not a divine gift but grows (or not) as the result of a learning process.

The history of number can best be understood as the development over many thousands of years of a special way of paying attention to, talking about and solving certain kinds of problem. These problems and solutions help human beings – the only creatures on the planet capable of thinking about mathematics – to overcome the limitations on their activities set by the social, physical and organic environment.

Mathematics has been called 'the language of the (exact) sciences'. Like other languages, it is constantly checked against experience, and altered to make it as simple and precise as possible. We begin with spoken numbers. Each number word points uniquely to one item or group of items, and excludes all others. The sequence of numbers precisely indicates the relations between smaller and larger numbers. Secondly, we have notation: that is, written symbols for each number and for the processes we carry out with them. Devising a suitable notation is a crucial step in developing a viable number system. A good notation reduces the amount of thinking needed to resolve number problems. It enables us, first, to make a clear abstract representation, in words and drawings, of each problem, and second, to devise an easy, almost mechanical routine (an algorithm) to deal with it.

Number and society

Numbers and calculation are evidence of the inventive genius of the human race. One purpose of this book is to show how different number systems arose in different societies, and how each helped to shape the society which devised it. Mathematical knowledge, whether related to grandiose architectural schemes, to the logistics of trade or war, to determination of the calendar or (as in ancient Israel) to the kind of reasoning used to solve tricky legal problems, is not a thing separate and of itself, but part of the totality of human interaction.

In the Western tradition, mathematicians and other scientists are sometimes regarded (not least by themselves) as a breed apart. This is true to the extent that they are a specialist group. But it does not imply that they are any more cut off from social influences, including the normal streams of opinion, ethics and inspiration, than any other specialist group: politicians, say, or plumbers. Nor does the 'star' explanation of human development (that each advance is the result of one 'star' individual's flash of inspiration and creative genius) fit the facts. In mathematics, such 'stars' as Pythagoras or Euclid often created more problems than they solved. The history of number shows that most advances were made by mathematicians working in the mainstream, relating their work to actual needs (for example, finding ways to predict the flooding of the Nile, to make correct tax assessments or to translate enemy codes in wartime).

Number in prehistoric times

Coleridge once remarked that if we could remember our nine months in the womb, we would find much of interest, far beyond anything that happened to us afterwards. Something of the sort is true of social history. If we knew the unwritten story of our past, especially the prehistoric part, its fascination would cut the history of kings and queens, wars and parliaments, down to proper size.

The last 200 years have seen an explosion of discovery about what might be called the 'foetal' period of human history – or at any rate that part of this period which follows the invention of writing. By deciphering ancient languages (for example, Egyptian hieroglyphics or Sumerian cuneiform writing on clay tablets, all decoded in the early 19th century), we have been able to enter not only into the public and religious life of ancient peoples, but into their very thought-processes. The general Western reader has had access to Hindu thought since the sacred Sanskrit scriptures were translated (by, for example, Max Müller) late in the 19th century. Ancient Chinese knowledge, available to foreigners for centuries but little considered, has been closely studied and assessed, by Joseph Needham in Cambridge, over the last 30 years.

However, if we seek to go back beyond the invention of writing (probably in Sumeria some 6000 years ago), we find ourselves in a pool of darkest night. The mysterious links of unrecorded history are hidden. We have no guide through the maze of non-literate cultures. Archaeology tells in detail of the material life of prehistoric families, but in the absence of a written record, we have virtually no access to their inner life and thought.

In particular, there are huge gaps in our knowledge of the origins of number concepts and of the transfer of ways of solving problems (algorithms) from one group of humans to another. As ever in such situations, where there is a problem but no evidence to speak of, people spring forward with this or that half-baked supernatural explanation. In this case, we are told that galactic time-travellers visited Planet Earth for just long enough to pass along some esoteric lore which it has taken our species all the time from then till now to understand – if we do understand it now.

Another suggestion, equally mysterious but better documented and less nonsensical, compares the spread of number-culture in prehistoric times to a known modern phenomenon, the children's 'bush telegraph', from which adults are normally excluded. Instances have been recorded by historians Peter and Iona Opie of how, in Britain, children's rhymes, riddles and comments on current affairs travel hundreds of miles in short order (sometimes days or even hours) along this network, whose existence is often as unsuspected by the young people who take part in it as by the adults who observe its results.

In the same way, scientific studies show that non-literate societies throughout the world, and throughout history, have worked from a

common base of values, superstition, folklore and calendar-linked custom
and behaviour which is unaffected by geographical, social and historical
boundaries. And if superstitious belief and ritual are transmitted in this
way, why should not factual, correct and even advanced scientific
information travel by similar unseen means?

The few insights that we have about prehistoric mathematical knowledge
come from cave paintings, a handful of domestic and ritual artefacts, and
buildings such as standing stones, graves, burial mounds and cairns. The
buildings by themselves point to the existence in prehistoric times of
considerable mathematical sophistication. As a minimum, the builders
must have had a 'feeling' for practical geometry, including an interest in
straight lines, circles and ellipses. They would also have needed to know, or
at least been able to use, the principle of the lever and the general physics of
massive bodies.

Two Scottish scientists, Professor Alexander Thom and his son of the
same name and title, studied and measured hundreds of cairns and standing
stones in Britain, especially in Scotland. Frequency (statistical) analysis
showed the use of two common units of measurement, the 'megalithic yard'
(equal to 2.72 feet) and 'megalithic inch' (one-fortieth of a megalithic yard).
The Thoms found that the 'yard', at least, was used to measure not only
length, but also the diameters of stone circles, some 67 per cent of which are
true circles based on it. They also showed that the megalith-builders
understood what we know as Pythagoras' theorem millennia before the
ancient Greeks, the ancient Chinese and the ancient Egyptians.

Our prehistoric ancestors around the world are also known to have
studied planetary movements, solar and lunar eclipses and star alignments.
Many 'siting lines', made of aligned pieces of rock and used as prehistoric
'telescopes' (zero magnification), have been discovered in North America.
Heavenly lore, and the priests, shamans and elders who interpreted it, were
central to the life of most societies. Pre-literate seafarers – as Thor
Heyerdahl and others have demonstrated – made maps using twigs and
sticks to represent ocean currents, took bearings on the stars and travelled
immense distances across open sea.

Civilisation and number

The next stage in our species' conquest of the environment, that of urban
civilisation, began about 6000 years ago and still exists today. It is marked
by a previously unknown phenomenon: the growth of centres of population, vast stores of wealth, enormous supplies of agricultural goods and an
overriding state apparatus. Development in such areas as commerce,

navigation, time measurement, urban planning and warfare, makes great demands on technologists and scholars. Such advanced mathematical skills as complex arithmetic, algebra, geometry and trigonometry become essential. Numbers and their manipulation become central concerns.

At about the same time, the invention of writing allowed sophisticated number systems to be developed and used, and the results recorded in forms which still survive, written on papyrus or paper, incised on clay tablets, painted or carved on monuments. This book details some of these systems, showing the interrelationship between each and the culture in which it arose. The record allows us, so to speak, to look over the shoulder of an Egyptian scribe as he calculates how to divide a day's supply of bread and beer among the workers at the Temple of Illahun. We can sympathise with a Chinese official faced with the task of organising a schedule of close encounters between the emperor and members of his harem – 1 empress, 3 consorts, 9 principal concubines, 27 assistant concubines and 81 slaves: a neat geometric progression (see page 66). We can oversee the plans for the construction of King Solomon's temple, compare ancient Greek ideas on number (to their disadvantage) with those of the Chinese (see pages 74–6; 53–4), or follow the thoughts on number-science of the Persian astronomer-poet Omar Khayyam (see page 149).

Such eavesdropping reveals a darker side. So far as the development of calculation is concerned, it cannot be overstated that the whole line of development lay outside Europe: in Sumeria, Babylon, China, India and the Arabian peninsula. Until the breakthrough by the Arabs in the scientific renaissance of the 7th–15th centuries (see page 137), Western Europe was a mathematical backwater. The reason was the baleful legacy of the Greeks. By treating numbers as semi-divine entities, nothing less than the building-blocks of creation itself, they removed them from everyday speculation and analysis into the realms of religion and philosophy. Mathematics became an esoteric subject, studied by a select few (Plato called them 'golden souls') and regarded with superstitious awe by the majority – a status which the science retained in Europe for two millenia, and the effects of which can still be felt today.

With the Arab scientific renaissance, Eastern number-knowledge, particularly that of India and Arabia, became accessible and respectable to Western scholars. From this point on, the march of mathematical knowledge was indeed guided, for the first time in history, by 'stars': individuals of genius such as Napier, Newton, Boole and Turing (to single out only British 'stars' in an international galaxy). The march also, inexorably, leads to the replacement of old, Greece-haunted ideas and systems by 'new' (actually ancient) notations and algorithms, and to the invention of the computer and the algorithms which make it possible. This device has not only transformed mathematics and science generally, but is the single piece of technology that best defines our age. It is the culmination of the story of number thus far in time.

In general, like most human activities, mathematics has had peaks and troughs, not following a chronological or geographical pattern but appearing to be more like random events, so to speak, in the continuum of human experience. This book focuses on moments that seem vitally important, or fascinating, or both at once. I have left out many 'big-name' discoverers, ideas and events. This is not because, say, Archimedean ballistics, the Aztec calendar or the theories of Gödel are insignificant. There are two reasons: first, that in the limited space available, they have yielded place to more central themes and issues, and second, that credit is often given to individuals for discoveries they failed to make themselves – national chauvinism being a compelling reason for continually falsifying the historical record in their favour.

1 2 3 4 5 6 7 8 9 10 11 12 13 14 15 16 17 18

THE LANGUAGE OF NUMBER

One's enough. Two's company. Three's a crowd.
Folk wisdom

Counting

Animals have an intuitive awareness of number. This means that they know, from experience, without analysis and immediately, the difference between a number of objects and a smaller number. In his book *The Natural History of Selborne* (1786), the naturalist Gilbert White tells how he secretly removed one egg each day from a plover's nest – and how the mother persisted in laying an extra egg every day to make up the original total. In more recent times, research has shown that animals – hens, for example – can be trained to distinguish between (what we call) odd and even numbers of food pieces.

This is, however, as far as it goes. Animals can only respond to a number situation when – as with eggs in the nest or food – it is connected to their species and survival needs. There is no transfer to other situations, or from concrete reality to the abstract notion of number. Animals can only 'count' when the objects are present and visible, and when the number is small (not more than five or six). In laboratory experiments, if you train an animal to 'count' using one kind of object, and then test it using another kind, it is quite unable to make the connection: the objects, not the numbers, are what interest it.

The reason for animals' inability to separate numbers from the concrete situation is that they are unable to think in the abstract at all – and even if they could, they have no language capable of communicating, or absorbing, such abstract ideas as 'six' or 'a herd'. A few pet-owners dispute this, claiming that their cats, dogs or horses understand words just as humans do. Experiments have shown that this is not so. A typical example, from the 19th century, warns us not to be so credulous. 'Clever Hans' was a horse which had been taught to 'think' and 'count', after several years of instruction by its devoted trainer, who was trying to make a point about horses being cleverer than humans. However, tests carried out by the German Psychological Society subsequently showed that, far from thinking out the questions it was asked (for example, working out addition sums and tapping the answers with its hoof), 'Hans' was in fact doing no more than responding to subliminal visual cues (to start tapping, and to stop) given unawares by its trainer.

Even human beings, for whom the abstract activity of counting seems the most natural thing in the world, find it incredibly difficult to learn. One of the discoveries of 20th-century scientists (made independently by Montessori, Piaget and Vygotsky) was that adults forget how gradual and time-consuming the process of learning to think in the abstract is: to make, for example, a connection between a series of apparently unrelated objects such as 'ships, and shoes, and sealing wax, and cabbages, and kings'. (To save anxiety, the connection is that they represent a set of 'miscellaneous objects'.)

What happens is that very young children learn from other people, in natural situations and in school, how to break a group of things up into its constituent units, and how to count them one by one, intoning the number sequence as they do so. This ability is built partly on the primary intuitions which human beings share with other animals, but mainly on what we learn from other humans: a special word for each number, for example. The child learns to think of numbers in the abstract (that is, without attaching them to objects or people), to think of them in the past and future, to deal with number aspects of faraway places he or she has never seen. Older human beings can even learn to think of imaginary numbers (such as infinity, or the square root of minus one), which no one can ever count. Such numbers have no physical meaning whatever until we give them one.

When Lewis Carroll talked of kings and cabbages, he was itemising a 'set': a collection of some kind. Provided that such a set is not too large, counting its individual components is no problem. We can say, for example, 'Here is a set of words which describe miscellaneous objects: one, ships; two, shoes; three, sealing wax; four, cabbages; five, kings'. Then we might continue, 'And we have just produced another set of five things, namely the number words one, two, three, four, five'. In creating these sets, both of five elements, we have set up (so to speak) a one-to-one correspondence:

one	two	three	four	five
ships	shoes	sealing wax	cabbages	kings

To make such a correspondence between a collection of (related or unrelated) objects or groups of objects and increasing numbers – in short, to count – we need a set of number words related to each other in certain ways. Most important, each must be uniquely different from all the others. They must start at 'one' and must follow each other in set order, one at a time, one after the other. Each must be associated with one thing only in the collection of things we are counting. The number linked to the last item in the set is the 'cardinality', or size, of the set. In the case above, the final cardinal number is five; our miscellaneous set consists of five kinds of items. Because it is now ordered or numbered, we can also give each item in it another (abstract) identification, that of its place in the order we have assigned to it. The first item is ships, the second shoes, the third sealing wax, and so on.

In short, we can use the set of number words, devoid of any concrete content or reference, to describe the miscellaneous set, and we can point to each object in the set, identifying it in a unique way. Most people learned these rules about counting before the age of five, and have probably never thought about the subject since. It still takes us about ten years of intensive schooling before we can follow this abstract description of the process. It is hardly surprising that so many bizarre explanations arose about where these abilities originated.

It took humankind many thousands of years to work out the rules or methods (algorithms) for counting. There was a time, not many centuries ago, when these skills were confined to an élite section of the population. In the modern age, few non-literate and non-numerate groups (what educated chauvinists used to refer to as 'primitive tribes') survive. But until about 1950, in remote corners of the world, there were groups of humans who were not much further advanced in their thinking about number than the 'one . . . two . . . many' implied by the folk saying that heads this chapter. When they made contact with 'civilised' traders, such people were at a disadvantage because they had no number words beyond three. They were forced back on setting up a concrete correspondence between the things counted and tallies by pebbles, parts of the body, notches on wood, or traded goods exchanged one by one for what they had to offer. Sir Francis Galton (Darwin's cousin) reported, as late as the 1880s, that a Damara tribesman, in Africa, exchanged sheep for tobacco by taking to twists of tobacco for the first sheep, then two for the second, then two for the third and so on. The process could not be hurried along without throwing him into total confusion.

In a survival situation, such as solitary confinement or shipwreck, prisoners or castaways tally days one at a time, recording each mark on a convenient surface. The 19th-century German poet Adalbert Stifter, looking forward to a meeting with his beloved, filled a bag with apples and ate one per day until the waiting period was over. As he wrote to her, 'When I wrote my last letter to you there were 21 – tomorrow there will only be 13. Finally, only one apple will be left, and when I have eaten that, I will shout for joy'.

By a natural extension of such tallying, instead of pebbles or notches or apples, various parts of the body could be touched to indicate different numbers. Everybody at some point counts on the fingers. Words are not really needed. It seems natural to count in groups of five, touching our fingers as we go. The next largest group is 10 (two handfuls). Taking our shoes off, we can progress to 20. In many languages, the origins of number words have to do with the body. 'Digit' comes from the Latin *digitus* ('finger'); the Indonesian *lima* ('five') also means 'hand'; the ancient Aztec *matlactli* means both 'two hands' and 'ten'; the word for 'whole person' in a great number of languages also stands for 'twenty' (ten fingers and ten toes). The dual number (a special way of saying two of something, preserved in many languages, for example 'pair' or 'twain' in English) testifies to the fact that we have two eyes, two ears, two 'noses' (i.e. nostrils) and so on.

Most number systems are recursive: that is, they go only to a certain point, for example ten in the decimal system, and then start again adding the words one to ten as a suffix to the base of the number system. For example, in English, 'eleven' and 'twelve' are Anglo-Saxon for 'one-more' and 'two-more'; 'three-ten' becomes 'thirteen', 'four-ten' becomes 'four-teen' and so on. After twenty ('two tens'), the units' place is moved to make the base number the important part, as in twenty-one, twenty-two and so on. Each of the European languages has similar conventions about numbers, which vary according to linguistic and social processes, about which nothing certain is known.

Where numbers are related to parts of the body, each number reference can, on occasion, be left as a gap in the verbal exchange, the gap being filled by a gesture. In 19th-century Papua, for example, Christian missionaries observed that one tribe numbered from one (the right little finger) through the other right-hand fingers, the wrist, elbow and shoulder to the right ear, right eye and left eye, through the face and so on down to the left little finger (twenty-two). But however elegant, this was not a proper number sequence. The system did not build on itself, using one concept to lead to others more complex, and had no logical organisation. The word *doro*, for example, meant any one of three fingers on the left hand or three on the right hand. One had to count through the whole system to discover which of the six possible meanings of *doro* (2; 3; 4; 19; 20; 21) was intended.

No such problems arose with the counting system described by Saint (formerly the Venerable) Bede (672–735), one of the most noted Christian scholars of the early Middle Ages. His interest in number arose in the context of the Roman Church's decree (Council of Nicaea, AD 325) declaring the method of calculating Easter. The Council proclaimed Easter to be the first Sunday after the first full moon of spring. (This ensured that Easter would never coincide with the Jewish Passover, which falls on the eve of the first full moon.) In the Roman Church calendar, the date of Easter establishes the dates of other movable feasts. On certain of these holy days all Catholics had to attend Mass or have a valid reason for not doing so, or else they were guilty of a mortal sin. In the case of Easter, the penalty was excommunication. Thus, exact calculation of the date affected many people's hope of comfort in this world and salvation in the next.

Bede set out ways of calculating Easter, and from there moved on to rationalise the whole calendar of world history. It was he, for example, who started the custom of dating events as being before or after the Incarnation of Christ (BC or AD). His *De computo vel loquela digitorum* ('On Counting and Speaking with the Fingers') is a complete exposition of how to represent all the numbers, from one to a million, by hand signals. This book was a prime mover in arithmetic in Europe for over a thousand years. Bede's system has two modern analogues: the 'tic-tac man' at horseraces, whose job it is to signal to his employer, by means of hand and body signals, the 'odds' offered by rival bookmakers; other 'tic-tac men' signal share prices in certain stock exchanges (New York, Tokyo, etc.) when they are busy and business is brisk.

The number sequence

The number sequence runs from one to the largest number anyone can imagine, or rather – since there is theoretically no end to the sequence – to infinity. Since about AD 1500, mathematicians have also been involved with 'negative numbers': that is, numbers less than zero. The number sequence has been taken as starting at zero and proceeding to infinity in both directions: there is a positive infinity and a negative infinity.

Understanding of this series developed very slowly. At first, zero and one were denied the status of numbers altogether. This was because Aristotle defined number as being an accumulation, or 'heap'. Since neither zero nor one constituted a 'heap', they could not be given the status of numbers. (In a similar way, though with less reason, some later writers on arithmetic also excluded two.) Like most 'rhetorical' distinctions of this kind (that is, to do with verbal quibbling rather than scientific fact), it made no difference in calculations, but limited the vision of experts of the day about the real nature of numbers and their functions in depicting space and time.

The history of number is embedded in the number word sequence as worked out in language. Names of the numbers for ten or less are among the oldest and most stable words in any language. This – and the remarkable similarity between them in related languages – suggests antiquity. Names for larger numbers, by contrast (100; 1000; 10,000 and so on) often seem to appear out of the blue, as it were, as if unrelated to anything else. This indicates a late borrowing of the word from another language. Similarities and differences in number names thus point to different stages in the development of the system.

To illustrate, let us consider the word 'googol'. This is a very large number indeed, standing for 1 followed by 100 zeros. It was coined fairly recently by an American scholar who had run out of prefixes to attach to '-illions' ('millions', 'billions', 'trillions', 'zillions' . . .). He is said to have coined 'googol' from the babblings of his infant nephew. It has now been borrowed by other languages. To a future linguist, its lack of relation to other number words and its occurrence in unrelated languages such as American English, Chinese, Hebrew and Russian will indicate that it is a late innovation.

At the opposite end of the size and time scales, the number names in West European languages (English, German, French, Italian, Spanish, Greek) form a well-defined and closely related group in terms of grammar, vocabulary, syntax and phonology. Their antiquity is shown not only by this relationship, but by their similarity to words for the same numbers in another, collateral group of languages, the Eastern European group which includes Polish, Czech, Russian, Lithuanian and Serbian. The table below,

English	French	German	Greek	Russian	Polish	Lithuanian
one	un	eins	heis	odin	jeden	vienas
two	deux	zwei	dyo	dva	dwa	du
three	trois	drei	treis	tri	trzy	trys
four	quatre	vier	tettare	chetyre	cztery	keturi
five	cinq	fünf	pente	pyat'	pyantz	penki
six	six	sechs	hex	shest'	sześć	sheshi
seven	sept	sieben	hepta	syem	siedem	septyni
eight	huit	acht	okto	vosyem'	osiem	ashtuoni
nine	neuf	neun	ennea	devyat'	dzyevyat	devyni
ten	dix	zehn	deka	desyat'	dzyesyat	deshimt
father	père	Vater	patir	otyetz	ojca	tevas
mother	mère	Mutter	matir	mat'	matka	motina

Note: ' stands for the Russian soft sign, which changes the sound of the ending.

giving a list of the words for one to ten in some of these languages, brings the point out clearly. (Words for 'father' and 'mother' are also given, to show similar relationships.)

Comparing other number words in both related and unrelated languages, we find that different nations have the same fundamental ways of dealing with plurality. It starts at birth. The infant, the eternal egoist, the I, 'numero uno', takes the first step. With great difficulty, in the course of time, the number two (that is, you) is identified. 'One, two, many' is how we apprehend the number system at this point. Then (at least according to Menninger) 'many' is replaced by the name of the next number in the sequence: three. The indefinite concept 'many' is brought into focus and given a clear identity as anything more than three. 'Many' moves along, and so do we. The system is now 'one, two, three, many'. And so on.

At this point, colloquial language may still reflect the earlier stage. For example, instead of saying, 'It's not very far', a Scot may say, 'It's just two steps'. Instead of saying, 'I met few people', an Italian may say, 'I met four cats'. This is the sense in which the Bible speaks of 'forty days', and of 'forty years' (as in the wilderness). Each phrase is intended to suggest a much larger number, the actual total being undetermined. Greeks talk of 'forty petals' in a rose; Sheherazade told stories for 'a thousand nights and one night'; as a very young child, I remember offering my mother 'a million kisses' in a number game we were playing. We can show the history of early (i.e. both infant and prehistoric) human experience with number on a kind of time line, as shown below.

---

1 (me)—— 2 (you)—— 3 (him or her)—— 4 (fingers)——5 (hand) —— 6—— 7——
8 (two hands, no thumbs)—— 9—— 10 (both hands) —— 11 —— 12—— 13 —— 14——
15 (and one foot) ——16—— 17—— 18 —— 19 —— 20 (one complete calculating human)

It remains to be said that despite the word 'googol', number words, like number systems, are seldom created by single individuals. They are social products, the result of group activities. They require social consensus to be viable. Scholars later write them down (and as a result are often credited with inventing them). But there is an *Alp* (old German: 'nightmare') of tradition which weighs on the minds of the living. We need to overcome the inert scholarly thinking which assumes that real life is what is preserved in libraries. It was not Archimedes, Bede, Boethius, Fibonacci or Newton who invented mathematics. It was ordinary men and women, unknown and uncredited, who made the fundamental advances. Scholars play an essential role in checking for consistency and applying intellectual discipline. But creativity, like morality, is something else again.

The first computer – two hands, ten fingers

The 'law of least effort', which motivates most human innovation, says in essence that the secret of success in human affairs is to mobilise your resources of energy, thought and procedures, only to the minimum extent necessary to achieve your purpose. William of Occam (14th century) put it in two pithy pieces of advice: 'It is futile to use more things to do the work when fewer will serve', and 'Do not multiply entities beyond necessity'. 'Occam's razor' (as these statements were nicknamed, because they sliced through the philosophical verbiage that passed for scholarship in his time) is now a basic principle in science, where it is known as the principle of economy.

A second, complementary principle, the law of chronic inertia, might be stated thus: 'Human beings will believe, or do, almost anything to prevent any change in their traditional routines'. This is really the law of least effort expressed in a different context. Both laws are exemplified throughout the history of number.

The number sequence discussed above (page 11) is formed by taking a starting point (one – but computer buffs should not forget zero) and then adding successive units until you have enough. Thus, we count from one to ten using the most convenient tools: our fingers. When these run out, we choose the next most available reckoners: toes (called in Gaelic 'the fingers of the feet'). These take us to 20 – at which point we make a breakthrough to a new system. We invent words to describe each tenth number: 'twenty' ('two tens'), 'thirty' ('three tens'), 'forty' ('four tens') and so on. This changes the rules. As we count using these words, we can forget the intermediate numbers. The system 'works' normally in terms of adding units. But we can now add in groups of ten. And something else has happened. 'Two tens', 'three tens' and so on are not just words in the number sequence: they describe the process of multiplication. They also show that it is a quick way to do addition. Instead of laboriously adding successive ones, we simply multiply (that is, in the decimal system, we add on bundles of ten units) to reach the next levels.

This insight also led to the discovery that the hands with their fingers could be used for more advanced work in calculation. Though dismissed by scholars as 'peasant arithmetic', finger calculation was an established method in office accountancy in Europe for many centuries, and is still used in parts of Eastern Europe. It even has a grand 'scholarly' counterpart: Bede's finger signals for numbers.

To give an example: this is how to multiply two numbers, each less than 10 – say 7 by 9 – without benefit of tables. Hold up both hands, palms facing you and fingers extended. (Thumbs count as fingers.) With the left hand, form the excess of the first number over five: in other words, bend down two fingers (since $7-5=2$). Do the same with the right hand for the second number. This time, bend down four fingers (since $9-5=4$). Next, add the bent fingers ($2+4=6$), and multiply this total by 10 ($6 \times 10=60$). Next, multiply the straight fingers ($3 \times 1=3$), and add the two totals ($60+3=63$). The answer is $7 \times 9=63$.

To multiply numbers larger than 10 – say 13×14 – the procedure is similar, except that we start by bending down fingers to show the excess over ten, not over five: left hand three fingers, right hand four fingers. Next, add them together ($3+4=7$) and multiply by 10 ($7 \times 10=70$). Multiply bent fingers ($3 \times 4=12$) and add the two results ($70+12=82$). Finally, add 100 ($82+100=182$). The answer is, $13 \times 14=182$.

How does the algorithm work? (It must be stressed that, in Bede's day, so far as most users were concerned, this was a non-question. If it worked, as it manifestly did, why try to explain?) To make it general, let us call the numbers to be multiplied A and B. The answer we are looking for is then $A \times B$. The algorithm and its working can now be tabulated as below.

To find the product of two numbers, A and B	*Example*
(i) When both A and B are less than 10	$A=7; B=9$
Finger bent: left $(A-5)$; right $(B-5)$	$7-5=2; 9-5=4$
Add bent fingers together	$2+4=6$
Their sum $=A+B-10$	
Multiply by 10	$6+10=60$
The tens figure is $10 \times (A+B-10)$	
Multiply straight fingers	$3+1=3$
Add results of the two multiplications	$60+3=63$
This gives the answer to $A \times B$	$7 \times 9=63$
(ii) When both A and B are greater than 10	$A=13; B=14$
Fingers bent: left $(A-10)$; right $(A-10)$	$13-10=3; 14-10=4$
Add bent fingers together	$3+4=7$
Their sum $=A+B-20$	
Multiply by 10	$7 \times 10=70$
The tens figure is $10 \times (A+B-20)$	
Multiply bent fingers	$3 \times 4=12$
The units figure is $(A-10) \times (B-10)$	
This is equal to $AB-10A-10B+100$	
Add results of the two multiplications	$70+12=82$
Add 100 to this total	$82+100=182$
This is	
$10 \times (A+B-20)+(AB-10A-10B+100)+100$	
This gives the answer to $A \times B$	$13 \times 14=182$

As the two examples (9×7; 13×14) show, this algorithm has one problem: it changes its form as we go up by fives. None the less, it is an excellent example of the way number problems throughout history have been simplified, often by an indirect approach – and also of the way numbers and number problems are subject to the laws of step-by-step logic, not mumbo-jumbo.

Written numbers

Numbers were spoken for hundreds of thousands of years before they were written down. And even written numbers show two quite different stages of development. Historically speaking, every 'sophisticated' set of written numbers, Babylonian, Egyptian, Chinese, Greek, Roman or Hindu, was preceded by or coexisted with a more 'primitive' set, used by ordinary people for barter and other sharing activities. The symbolic representation of numbers begins in an unwritten way, using such methods as body signals and pointing at the objects to be counted, finger tallies, marks on the ground or in the sand, arrays of pebbles, and piles of shells or beads. In the few cases where ordinary people made permanent records of counts, their 'numbers' usually followed the same kind of aggregative method: just as you added one more pebble to the pile for each new item added to the number, so you might add another dot or line or tally to the written account.

A good example of such 'primitive' methods of 'writing' numbers down is tally sticks: that is, pieces of wood with marks cut in them. These were used in peasant communities for millennia. They were regarded by the state as legal documents, and sometimes even served state purposes. From the 1300s to as recently as 1828, for example, the British exchequer made tax demands, and gave tax receipts, in the form of tally sticks. When the system was abandoned, an enormous pile of sticks remained, stored in the vaults of London's Houses of Parliament. In 1834 it was decided to destroy them by burning. During this operation the buildings were accidentally set on fire and burned to the ground. (Turner painted two remarkable pictures of the conflagration.)

The marks whittled on tally sticks were not standard symbols such as those we use today. They were idiosyncratic and personal to the individual keeping the tally. This was to identify who had used the tally stick as a bill or receipt in disputed cases. Different marks might even be made by the same individual when numbering different things. For example, 'ten' referring to cows was often a different symbol from the 'ten' used for loads of fodder or containers of milk. The tally-makers in such cases had not sorted out that

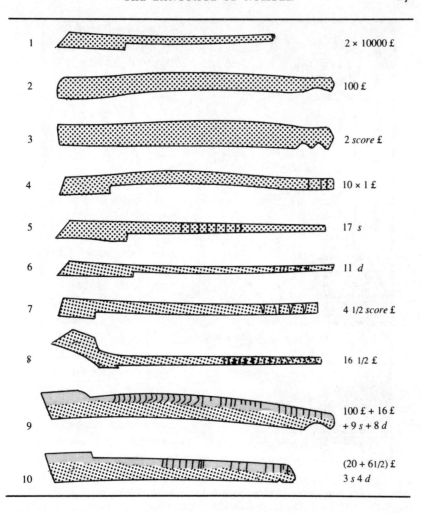

1	2 × 10000 £
2	100 £
3	2 *score* £
4	10 × 1 £
5	17 *s*
6	11 *d*
7	4 1/2 *score* £
8	16 1/2 £
9	100 £ + 16 £ +9 *s* + 8 *d*
10	(20 + 61/2) £ 3 *s* 4 *d*

different things could occur in equal numbers, and that any association between the numeral and the thing it referred to was merely temporary. (This may seem incredible, until we recall that number was until recently surrounded by a semi-magical aura. As few as 100 years ago, Polish peasants religiously did not mix money intended for a daughter's dowry with money set aside for buying land. The reason was not prudence but because they regarded the money itself as different. The two sums were like heaps of different seeds.)

Simple records, such as those on tally sticks, served most of the needs of ordinary life and trade. But for more solemn purposes, such as recording the triumphs of ruler or state, drawing up calendars or codifying laws and taxes, written numbers were required which were simultaneously more

Illustration from Boethius' *The Consolations of Philosophy*. The philosopher
(left) is using Arab numerals; the merchant (right) is using a counting-board to
reckon 'on the lines'. On the goddess' dress are two geometric series: 1, 2, 4, 8
and 1, 3, 9, 27. The idea which the picture seems to represent, of competition
between the two systems, could hardly be further from the truth of history.

sophisticated to use and grander to look at. In many societies, a division
grew up between people who used calculation for practical purposes and
those who used numbers for ritual or state business. The latter groups
quickly became a self-proclaimed élite, and the written numbers they used
were regarded as special, even sacred. In some cultures, they were claimed
as the invention of the gods themselves. Confrontation (or perhaps rather,
competiton) between the two groups began in earliest times, and continued
far longer than some of the great civilisations of which it was characteristic.
In a 14th-century manuscript of Boethius' *The Consolations of Philosophy*,
for example, there is a famous illustration showing two 'mathematicians', a

Arabic	Egyptian	Babylonian	Mayan	Chinese	Greek	Roman
1						I
2						II
3						III
4						IV
5						V
6						VI
7						VII
8						VIII
9						IX
10						X
26						XXVI
723						DCCXXIII
11,578						TMDLXXVIII

merchant with his abacus and a Pythagorean philosopher with his 'sacred numbers', indulging in a competition supervised by the goddess of number. The participants in this picture would have been easily recognised by all who read the book – and no reader, even at this late date, would have seen anything strange in either the situation shown or the dichotomy it symbolised. This was seen as a war between rival 'unions' and as a very serious business.

Because 'sophisticated', or élitist, systems of written numbers were divorced for so long from everyday calculation, many of them could steer well clear of practicality with no penalty to the users. They were the tools not of science but of rhetoric; their manipulation was concerned less with measurement (in the widest sense) than with philosophical speculation, often about the behaviour and purposes of supernatural beings. Symbols that would not have survived for five minutes in the rough-and-tumble of ordinary life lasted for centuries, and inhibited scientific advance in societies which were in other respects – government, commercial enterprise, the arts – among the most glittering 'civilisations' the world has known. Or so it is said.

The problem can be quickly stated. For calculation, simple number symbols are easier to use than complicated ones. Thus, the 'simpler' the written numbers, the more 'advanced' the arithmetical work they permit. For advanced calculation, numbers based on letters of the alphabet, or using the initial letters of such words as *mille* (Latin: 'one thousand'), or based on accumulations of symbols, as in Egyptian numbers on stone carvings, are awkward at best and a crippling handicap at worst. They can be confused with words, with each other or with random marks on the material written on (such as chippings in stone or blobs on papyrus). The table on page 19 makes the point – and also shows the superiority of the Hindu/Arabic numeral system which led to the one now in worldwise use.

THE AMERINDIANS AND NUMBER

Blood and cruelty are the foundation of all good things.
Hernando de Soto, Spanish conquistador

LONG before the continent of America was invaded by Europeans, it was already inhabited by over 500 different native groups: some small, others bigger federations, others large enough to be counted as nations. The variety of languages and traditions was as great as in Europe, India or Asia. Where the Amerindians (American Indians) originally came from has long been discussed; the current consensus is that they crossed the Bering Straits from East Asia, possibly at the end of the last Ice Age (10,000 years ago).

In general, Amerindian groups remained at the level of what European scholars now call the New Stone Age: that is, they made weapons of worked and polished stone. Only in a few places did they advance beyond nomadic savagery. The Incas of Peru had a complex class structure but no written language. The Aztecs of Mexico and Maya of Yucatan, two quite separate peoples, each had a developed civilisation and a form of picture writing. All three peoples, Incas, Aztecs and Maya, are perhaps best described as belonging to the Bronze Age, later in time but at roughly the same cultural level as the Mycenaeans in Europe.

Apart from the monumental temples and pyramids of Central and South America, and a few relics and folk traditions from further North, remains of this rich Amerindian culture are sparse. The genocide commited by the Spaniards against Aztecs, Incas and Maya was equalled only by the wars launched later by others against the native peoples of North America. None the less, Amerindian culture covers a remarkably wide spectrum of human progress, from the primitive drawings and number lore of North American tribes, through Inca and Aztec artefacts, to the picture language and complex number system of the Maya. This chapter looks at the whole range of Amerindian culture, with particular emphasis on North American concepts of number and the calendar, and at the Incas. Chapter 9 is devoted

to the Maya, whose understanding of number was far in advance of
anything else contemporary on the continent of Europe.

Amerindians of the North

The native peoples of what are now Canada, the United States and Central
America were diverse in character, with many tribal groups and a wide
spread of languages and interrelationships. Some groups were nomadic
hunters and fishers. Others engaged in agriculture or herded animals. There
were no horses until the Spaniards imported them. (This is in spite of the
fact that the horse originated and evolved in North America, some time
before *Homo sapiens*.) Goods and effects were transported by pack dogs,
sometimes pulling a sled or a *travois* (a wedge-shaped structure made from
two branches tied together, forming a kind of wheel-less cart: the wheel was
introduced by later immigrants from Europe).

Amerindian life was oriented towards hunting and tribal war. Religion
and culture were based on a spiritual conception of reality, on the idea of
harmony, if not between human and human, face to face, then certainly
between human beings, nature and the supernatural. Rather than being
analytical or empirical thinkers, the people were intuitive mystics. As a
result they made very few, if any, advances in scientific knowledge or in
number. They could count, but had no concept of abstract calculation, no
numerals, and no way to pass on their results except orally or by tallies
scratched on the ground.

Achievement in tribal wars and skirmishes was rewarded in several
symbolic ways. The number of times each brave led a war-party and the
number of enemies he killed were credited by special signs on his war tunic,
headdress or leggings. The number, position and slope of eagle feathers in
the headdress, their shapes, cuts and markings, signalled the wearer's
military prowess. Feathers were also used to decorate bows, arrows and

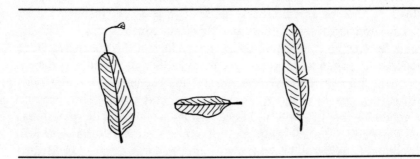

other weapons. There were tally marks on the war tunic, and tattoo marks
and war paint on face and body. Late in their history, warriors from some
native groups even carried scalps of their victims – a revolting habit
possibly learned from white bounty-hunters. All these things could be 'read'
by the initiated in similar fashion to the medals, good-conduct and long-
service stripes and badges of the modern career soldier.

Recording the exact number of such credits was about the greatest strain
ever placed on a warrior's arithmetic. Similar symbols were used in other
situations, none more sophisticated than were strictly needed by the
occasion. The two illustrated below are typical. In the first, the offer is made
to trade four animal skins (one buffalo and three others) for a rifle and 25
cartridges.

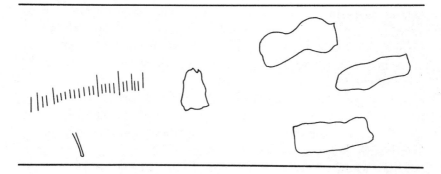

The second message records an Ojibway attack on a Sioux encampment of
tents. It shows (first) braves spying out the land and (second) the same
braves moving to the attack in single file, 'Indian style'.

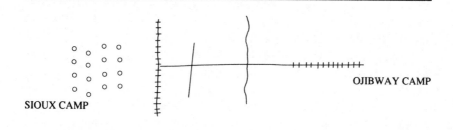

In summary, native peoples of North America, until and in some cases
beyond the arrival of white people (from whom they learned new methods
of counting, just as they discovered the uses of horses, whiskey and

gunpowder), counted on their fingers, and used objects such as pieces of wood and pebbles as counters. They recorded the results of their counting in tally marks. One or two tribes, for example the Blackfoot, used the *quipu* (see page 26) for tally purposes. But there is no record of any operation that goes beyond mere counting by enumeration. Lack of notation, even of numerals, made it impossible to refine the concept of number. The level of skill was adequate for addition to 20 (fingers and toes). It was perhaps just strong enough to support the slightly more complex notion of subtraction. But multiplication and division, if they existed – there is no evidence that they did – were probably specialist activities, practised only by a few.

The calendar

In confronting and solving a few key problems, human beings employ 'tools' to discover and understand reality better. Four activities in particular have been crucial for the development of number: the invention of money (especially metal coins or shells), the invention of the calendar, the measurement of time, and the use of weights and measures. The number problems that arose in connection with these activities, and relations between the various solutions found, led to major advances not only in the specialist area of calculation but in the exact sciences in general.

In North America, the absence of metal money (or other objects identified as a medium of exchange) and of a calendar (in any but the crudest sense) were not only main symptoms of the backward state of number, but also causes of it. The reason why the Amerindians were so handicapped was that, for millennia, there was an absence of any dynamic of cultural change. As the cultural anthropologist Malinowski has pointed out, the force essential for change is generated by an intrusive culture that violently substitutes new and exciting values for the old ones, or revitalises them in the way 'new blood' invigorates an effete familial strain. Such an intrusive culture will normally be at a higher level of knowledge and practical activity than the indigenous culture. (The Incas, Maya and Aztecs are rare exceptions to Malinowski's rules: in each case, it could be argued, the incoming culture was more primitive than the culture it encountered.) New technologies also help – for example, forms of rapid transport and coinage (a common medium of exchange between interacting cultures). These, and most other generators of change, were missing in America until the whites arrived.

Time, like money, can be divided into standard units, large and small. That these units can be used to measure the passage of time is the first discovery on the way to the concept of a calendar. Second is the insight that natural time-units can be established by study of the movements of star constellations and heavenly bodies, especially the Sun, Moon and (perhaps) Venus.

The alternation of night and day, seasonal variations of climate, the effects of changing patterns of weather, and the renewal and survival of life are the things by which we monitor and regulate our lives. Body needs too have a cyclical rhythm – the need for movement to avoid cramp, the longing for food or sleep, the sexual urge. Progress towards a calendar involves refining such rough measures, replacing them with better (because equal) units. Reference to an objective standard (the appearance of designated heavenly bodies, for example) in place of a subjective one (for example, hunger or fatigue) is one such advance. But no North American indigenous people worked out a 'true' calendar: that is, a single system of reference tying the cyclical nature of natural events to a scientific formula that measured the passing of time and allowed the reasonably certain prediction of future events. In North America, we see only the earliest glimmerings of such a process. The day was recognised as a unit, but there was no attempt to distinguish between days by naming them, or by recognising any succession or sequence. Except in a single instance, an account of tribespeople who died during one winter of disease, it was not recognised that days could be combined into larger standard units such as weeks, months and years, or broken down into such smaller standard units as hours and minutes. The day began at sunrise and finished at sunset. That its length varied with the season must have been known, but there was no attempt to measure the differences.

Even living life one day at a time, so to speak, it is possible, without creating fixed units, to take some steps towards organisation. A few groups, for example, used bundles of sticks to prepare for an event in the future. The known number of days was counted out stick by stick, then one stick was removed daily until they were all used up. Other groups made *ad hoc* tallies of days as notches on wood. Longer periods were counted in the same haphazard way: there was no organisation, no evidence of anything but reactive thought. One way used to talk of event in the past, for example, was in terms of a particular season or event: 'ten winters ago', say or 'seven moons ago'.

The Incas

The Incas arrived in Central and South America in about the 13th century of the Christian era. By 1500, under their leader Manco Capac, they had established an empire in an area now covered by parts of modern Peru,

Uruguay, Chile and Nicaragua. At its height, the Inca empire, ruled from its capital Cuzco ('navel') in the High Andes, covered a million square miles of territory.

In spite of their lack of a written language, the Incas set up a system of taxation and government of the greatest efficiency. Records were kept by means of *quipus*. A *quipu* (the Inca word for 'knot') consists of several lengths of cord tied together, or to a cross-piece. The cords can be of different colours, and knots are tied at intervals in each of them. The *quipu* was once widely distributed: for example, in 19th-century Africa, Bangala hunters made a knot for each deer or elephant killed. The ancient Chinese *Book of Changes* and the *Tao Te Ching* both refer to an older and better form of government 'by cords'. Tax collectors in the reign of the emperor Shen Nung (28th century BC) are known to have given *quipus* as official receipts. The *quipu* was also used as a tax record in ancient Greece. At the other end of the time line, the *quipu* was used until the 1830s in Hawaii: a society, like that of the Incas, with no written language. (Until recently, indeed, Peruvian shepherds used *quipus* to record the size of their herds. Each cord, coded by colour, stood for a different kind of animal. Rams, ewes, lambs, goats and kids were shown as knots on a white cord. Cattle were shown on green cord; bulls, mulch cows, barren cows and calves were counted as separate groups.)

Despite such hints, it is not completely clear how the ancient Incas used *quipus*: apart from the instruments themselves, no evidence survives. The *quipu* was almost certainly a memory device, maybe used for simple totalling operations. The principle was probably a sophisticated version of our tying a knot in a handkerchief to remind us of something. In the *quipu* there were many knots in sets, one set to a cord. They stood for the numbers of each thing to be remembered, recorded or calculated. The *quipu* was in the care of an official called a *camoyoc* ('rememberer'): a civil servant, a kind of animated file. The record consisted of two things: the *quipu* itself and the *camoyoc*'s verbal report.

It may be possible to shed some light on how such a record might work, by an ancient Tanzanian story of a man about to set out on a trip. Before he left, he took a piece of cord and tied eleven knots in it, equal distances apart. Then he said to his wife, touching each knot in turn, 'This knot is today when I'm starting. Tomorrow I'll be on the road, and I'll be walking the whole of the next day and the next. But here', he said, indicating the fourth knot, 'I'll reach the end of my journey. I'll stay here on the fifth and sixth days, and start home on the seventh. Untie a knot each day, and on the tenth cook food for me – for see, this is the eleventh day, when I'll come back'.

The Inca number system (perhaps taken over from one or other of the peoples they conquered) had a base of ten, and a minor base of five. For example, order in the empire was maintained by a system of intensive spying; to facilitate this the entire population was divided up into a hierarchy of groups, each with an appointed leader. The groups were of standard sizes – 10, 50, 100, 500, 1000, 5000 and 10,000. The census for each region, and records of the military supplies it needed, were kept by *quipu*.

For all such purposes, the Incas used a number system based on position. *Quipu* 'numerals' included a zero: a blank space on the cord. In *quipus* with coloured cords, the colours were probably used for classification, for example to indicate place value. Some strands represented units, others tens, others hundreds, and so on. As five places was the normal limit in counting, the *quipu* could represent any number between zero and 100,000. The total sum represented on the *quipu* might be given on a separate cord. (There is no certainty about any of this. I have speculated about Inca practice, based on what is known of how the *quipu* was used in other places.)

An alternative theory is that the colours were used for inventory, as by the Incas' Peruvian descendants in recent times. Different classes of things were shown by different colours. Tax records, land grants, economic production, religious ceremonies (especially when involving human sacrifice) and all matters of military information were almost certainly kept on *quipu* file in Cuzco. Leading state servants such as judges, army officers and heads of groups, and regional officials as lowly as village chiefs, probably also had such material 'on file'. The personnel needed to maintain the system must have been enormous.

To us today, conditioned by centuries of dependence on written documents *not* to remember things by heart, it may seem incredible that *quipus* could be used for keeping records in this way. But anthropologists and cultural historians report many 'extraordinary' feats of memory by illiterate or pre-literate people. In pre-Homeric Greece, it was usual for bards to recite whole sagas from memory (this was the origin of the *Iliad* and *Odyssey*). The whole of the Hindu *Vedas*, about three times as long as the Bible, was handed down by word of mouth. Written communication undermines oral memory. Inca calculation was done on fingers, by pebbles or notches. The *quipu* knots were reminders of totals. In the case of speeches and verbal reports, the knots might record headings and sub-headings in a long and detailed statement, such as the speech of an ambassador. The *quipu* was also used in school, to assist number work and the learning of such subjects as Inca history. The *quipu* was a mnemonic device, a kind of personal shorthand. It is like the speech of the Tanzanian husband to his wife. Looking at the *quipu* she would recall the details of his trip.

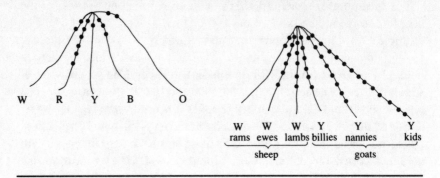

Using the *quipu* system, Inca state records were kept so accurately that 'not so much as a pair of sandals' could go astray. The officials kept in daily touch with all parts of the empire. They used a system of state couriers called *chasquis*, who ran on the royal road (the 'highway of the Sun') in relays. They bore *quipus* (and, no doubt, the accompanying oral messages) to the capital, returning to base with the official response. They could deliver *quipus* from Quito to Cuzco (about 1230 miles) in five days, and from Cuzco to Lake Titaca (about 750 miles) in three days. It was a matter of organisation. On the highway, post-stations were regularly spaced, every $4\frac{1}{2}$ miles. Each *chasqui* in the relay would run at top speed from one station to the next, at heights ranging from 10,000 to 15,000 feet above sea level. (Modern descendants of the Incas have performed similar feats, after training for them.)

One sinister application of the Inca *quipu* system remains to be mentioned. Recent research has shown that, contrary to earlier opinion (which suggested that human sacrifice was a feature of the civilisation of the Aztecs but not the Incas), yearly quotas of adults and children were ritually slaughtered, both in Cuzco and in local villages. The whole thing was planned as a kind of *quipu* simulation: as it were, a cord model of the empire. The cords were spread out to represent roads radiating from the capital, and the knots then represented quotas for sacrifice from the areas of population they covered.

So far as mathematics is concerned, the chief interest of the *quipu* was that it used the decimal system (to base ten). Its use by the Incas was only one instance of its widespread appearance in the arithmetic of pre-literate communities throughout the world. There may, indeed, have been a universal '*quipu* stage' in the history of number, replacing the earlier tally-mark system and preceding such things as the abacus calculations of ancient Greece, ancient Rome and China.

1 2 3 4 5 6 7 8 9 10 11 12 13 14 15 16 17 18

SUMERIA AND BABYLON

Everything has been said before by someone.
Alfred North Whitehead

IT was not until the first half of the 19th century that excavations in the Middle East revealed that a number of highly developed civilisations had existed there for at least 4000 years before the Christian era. About 150 miles north-west of the Persian Gulf, the ancient culture of Sumeria, whose existence had been unknown for millennia, was uncovered from the desert sand. Several thriving cities once occupied this area, including Akkad (capital of the almost-legendary king Sargon) and Ur (from where Abraham is believed to have migrated west in the 22nd century BC). Sumerian law and literature influenced several nations in Bible lands, not least the Jews. The Assyrians, Hittites and Babylonians (see below) adopted the Sumerian system of writing and many other aspects of their culture.

Rather more is known of ancient Babylonia than of Sumeria. The Babylonians rose to power nearly 4000 years ago, overwhelming the Sumerians and engulfing their empire. They ruled a huge area of what is now Syria, Iraq and Jordan, and built magnificent cities, notably Babylon, ruled by King Nebuchadnezzar I in the 12th century BC. Their empire survived until the 7th century BC, when it in turn was engulfed by the Assyrians.

Sumerian letters and numerals

The rise of cities radically changed people's culture and the way they organised their lives. The move to settled farming led to a surplus (very different from the hand-to-mouth self-sufficiency of nomadic peoples), and surplus led to wealth. A new social class came into being: landowners, who rented to others, taking a share of the product. Trade with other nations led to further division of labour. Crafts and industries such as pottery, leather

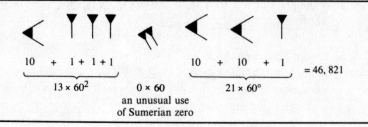

$$10 \quad + \quad 1 + 1 + 1 \qquad\qquad\qquad 10 \quad + \quad 10 \quad + \quad 1$$

$$\underbrace{\qquad\qquad\qquad}_{13 \times 60^2} \qquad \underbrace{\quad}_{\substack{0 \times 60 \\ \text{an unusual use} \\ \text{of Sumerian zero}}} \qquad \underbrace{\qquad\qquad\qquad}_{21 \times 60^\circ} \quad = 46,821$$

and metal-working were developed to a high level. Merchants set up markets and shops to sell goods and materials, both their own and those imported from elsewhere.

All these developments took place in Sumeria, perhaps for the first time in world history, in the 4th–2nd millennia BC. To serve the needs of their ever-growing populations, the Sumerians invented writing, using symbols to record both words and numbers. Their medium was clay, incised with a stylus and then allowed to harden. The stylus, a pencil-shaped cylinder with an oblique cut at one end, made three kinds of marks. A circle could be formed by pressing the round end of the cylinder into the clay, a semi-ellipse by laying the sharp end of the stylus at an angle on the clay, and a semi-ellipse with a straight line extension by pressing the whole cylinder into the clay. The Sumerians did not invent an alphabet – this came from the Phoenicians at a later date – instead, they represented the syllables of the language by different patterns in the clay. At different times, the number of words recorded by means of these three elements ranged from a thousand in the early days of writing to about four hundred in the later periods. (There was a general thinning of the written vocabulary.)

As with words, different stylus-patterns represented different numerals. A place system was used: that is, the position of each numeral in the number indicated its value. Apart from the numerals, we still know little about Sumerian arithmetic. The number system itself (sexagesimal) was very simple. There were only two numerals, 1 and 10. The base was 60. In other words, reading from right to left, the place-value went up by sixties, instead of by tens as in the decimal system. For example, our modern (Arabic; decimal) system tells us that the expression 315 means (3×100) $+ (1 \times 10) + (5 \times 1)$: the place of each numeral gives its value, moving from left (larger numbers) to right (smaller numbers). The Sumerian expression 3, 1, 5 means $(3 \times 60) + (1 \times 1) + \frac{5}{60}$, that is $181\frac{1}{12}$. (I have 'translated' cuneiform marks into Arabic numerals, but the principle is clear. The conventions, borrowed from Neugebauer (a German scholar who specialised in pre-Greek number systems), are that a comma separates each place from the following number; a semicolon marks the break between the whole-number part and the fractional part, as in 3;45,30, which stands for $3 + \frac{45}{60} + \frac{30}{60 \times 60}$ ^{or} $3\frac{91}{120}$.)

The Sumerian place-system, the first ever invented, was deficient in one respect only: there was no generally used sign for zero. This meant that you needed to have some context for the number before its size could be decided. It is easy to see the difficulty, if one imagines the same lack in our modern system – that when we are given a number, we are not told whether it is a fraction or a whole number, or where the decimal point should be. For example, taking a shopping list to buy groceries for someone else, we might read 'Beer: 1'. Assuming that our number system has no zero, we are faced with the problem: Does this mean one, ten or one hundred cans? Does it mean one-tenth or one-hundredth of a can?

We can narrow the problem down by common sense (eliminating the fractions, for a start). If we know that the beer is for lunch, say, we might guess that only one can may be meant. If it is for a party, then 10 cans may not be enough, but 100 will probably be too many. In the end we may settle for 6 (that is, one small pack) with the proviso that we can collect more before closing time. The area of uncertainty is reduced by knowledge of the context, but the system is unsatisfactory, even for such simple tasks as buying beer, and virtually useless for more sophisticated calculations.

This ambiguity is the one serious disadvantage of the Sumerian system. Otherwise it is superior to most systems used since then, at least until the metric system was established some 190 years ago. There were four main reasons for this superiority: first, the place-value concept; second, the extension of the numerical base of 60 to include fractions; third, the fact that there were only two numerals, for one and for ten, and last, the fact that the base of 60 was used also for weights and measures.

In other words, fractions were on exactly the same basis (a base of 60) as ordinary numbers. Thus, if there were fractions, no denominator was needed: it was always understood to be one. (This particular advantage was recaptured only in 1585 when decimal fractions were invented by Simon Stevin of Bruges.)

The advantage of using the same base for weights and measures was that the rules were the same as for fractions. The divisors for conversion of one weight to the next lowest level were, like the denominators, standard.

A main advantage of the sexagesimal (that is, where numbers are on the base 60) over the decimal system is that base 60 has many more factors than base ten in the decimal system. The factors of 60 are 1, 2, 3, 4, 5, 6, 10, 12, 15, 20, 30 (sexagesimal) as against only 1, 2 and 5 (decimal). This means that fractions and weights and measures can be expressed with great precision in the sexagesimal mode. There are relatively few repeating fractions. One-third is given as 0; 20; one-seventh is 0; 8, 34, 1028, 112000.

These points can be illustrated by showing the same measurement in different number systems (using Arabic figures for convenience). We measure a plank of wood; its length is 3 yards, 2 feet, 5 inches, and $\frac{3}{10}$ of an inch.

Yards: $3 + \frac{2}{3} + \frac{5}{36} + \frac{3}{360} = 3\frac{293}{360}$ yards
Metres: $3 + \frac{4}{10} + \frac{8}{100} + \frac{7}{1000} = 3.487$ metres
Sumerian: 3; 48, 50, that is, $3 + \frac{48}{60} + \frac{50}{60 \times 60} = 3\frac{2930}{3600}$ yards

Babylonian number skills

When the Babylonians engulfed the Sumerian empire in the 20th century BC, they took over the entire Sumerian heritage: script, literature, astronomy, laws, arithmetic, merely adapting these, as required, from the Sumerian language (which was unique, and related to no other in the area). Adopting Sumerian writing and number gave them a head-start in both science and mathematics. Their only innovation (appropriately, for a vast trading empire) seems to have been the system of weights and measures, which remained standard in the Middle East for 20 centuries. Its basic unit was the shekel (about 10 grams: that is, about quarter of an ounce). Larger units were the mina and talent. These were also units of coinage.

Far more evidence has been unearthed, and deciphered, about Babylonian trade and commerce than about the Sumerians. Hundreds of thousands of clay tablets survive; a large proportion, however, is crumbling to dust, year by year, undeciphered. These tablets relate to business activities: inventories, bills, receipts and deeds of sale. They far outnumber such other writings as royal decrees or religious texts.

From the mathematical point of view, one of the most interesting discoveries from these tablets is that the Babylonians used 'tables' for a great number of procedures: multiplication, division, fractions, square and cube roots, and much more. This made arithmetic almost a mechanical operation, just a matter of looking up tables. To illustrate, here is a problem from the 2nd millennium BC. It preserves the flavour of the work-books used by children at school at this time.

I multiplied the length and the width, to find the area. . . . The answer was 3,2 . . . . [that is, $(3 \times 60) + (2 \times 1) = 182$] . . . Then I added the length and width together; the answer was 27. Find the length, width and area.

Faced with this problem, the child (with luck) would at once recognise it as an 'area-of-the-field' routine. He or she would begin by checking if it referred to a square field. The square-root table would show that the nearest squares to 182 are 13^2 (169: too small) and 14^2 (196: too big). Possibly, therefore, the field was oblong, with sides 13 and 14. The multiplication table would show that 13×14 equals 182: correct so far. Length plus width $(13 + 14)$ equals 27: also correct. Everything tallies. The answer to the problem is length 14, breadth 13 and area 182.

The difference between a Babylonian and a modern schoolchild would be that the modern pupil would not have ready-reckoner tables to consult. Instead, he or she would follow our own 'area-of-the-field' routine. We have been trained to write down two equations (because we have two unknowns). The equations are: If L stands for the length and W the breadth, then:

$$L \times W = 182 \qquad \text{equation (i)}$$
$$L + W = 27 \qquad \text{equation (ii)}$$

We would solve the two equations by putting $27 - W$ in place of L in equation (i). We then have a quadratic equation in L which we could solve easily. The Babylonian method of attack is both simpler and less abstract: more closely in touch with the practicalities of the situation. It is 'rhetorical' algebra, not using letters for the unknowns but allowing us to think through the detail of the operation in words. The method (looking up the relevant tables and testing various approximate answers) works just as easily with fractions as with whole numbers, as can be seen in the example in the next section.

Babylonian algebra

Instead of the current method of using x, y and z for the unknown values to be found, and a, b and c for coefficients (the modern system introduced by François Viete, in 1591), Babylonian rhetorical algebra spoke of a 'side' as the unknown and a 'square' for the power of two. If there were two unknowns, they were called 'length' and 'breadth'; their product was the 'area'. If there were three unknowns, they were referred to as 'length', 'breadth' and 'width'; their product was the 'volume'. Once again, these terms are evidence for the close connection in early mathematics between theory and practical affairs.

Babylonian mathematicians were familiar with linear and quadratic equations, square roots and cubic equations. These were first reduced to a standard form. For example, there were six types of quadratic equations, each of which had a standard method of solution. A typical problem reads:

I subtracted the side from the square; the result is 14, 30. Find the side!

Realising that this was a second-degree equation (a quadratic), the pupil would impress it on wet clay, in rhetorical form, as follows:

> square diminished by side is 14, 30 units.
> (We would write this as $x^2 - x = 870$.)

This was one of the six standard forms. The student would then look up the instructions for solving this type of second-degree equation, as shown below.

For the problem 'Squares and sides yield a number':
Write down the number of sides (ignore the minus!): 1
Write down half of this: it is $\frac{1}{2}$
Then square it: $\frac{1}{4}$
Add this to 'square less side'
The number 'square less side', along with $\frac{1}{4}$, makes 870 with $\frac{1}{4}$
Look up square root table under $870\frac{1}{4}$: answer $29\frac{1}{2}$
We have thus shown that side less $\frac{1}{2}$ is the same as $29\frac{1}{2}$
So side must be 30. This is the answer.

This is exactly how we would solve this quadratic today, except that we would replace the words by letters such as s (for 'side'). The algorithm is known as 'completing the square', and is a standard method which can cope with any kind of quadratic. We learn the 'formula', and then it is only necessary to put the equation in standard form and substitute the coefficients (values of the specific problem to be solved) in the formula.

$s^2 - s = 870$
$s^2 - s + \frac{1}{4} = 870\frac{1}{4}$
Take the square root of both sides of the equation.
$(s - \frac{1}{2})^2 = 870\frac{1}{4}$
$(s - \frac{1}{2}) = 29\frac{1}{2}$
Add $\frac{1}{2}$ to both sides.
Answer: $s = 30$.

The Babylonians could also deal with number series, both arithmetic and geometric. They were interested in what have been wrongly called Pythagorean numbers. These are sets of three numbers where the sum of the squares of two of them is equal to the square of the third. Such numbers represent the sides of a right-angled triangle (see pages 80; 117). The Babylonians could also work out the area of a triangle from the length of the sides. They knew the formula for the volume of a pyramid, as well as for other geometrical shapes.

The Babylonian calendar and the beginnings of astrology

The Babylonians thought that the seasons were regulated by the gods, in the same way as different agricultural procedures were suited to each season. It was a priestly task to work out the royal or official calendar so that prayers could be said, and the correct rites performed to the right god at the right time.

The problems of survival that they faced were different from those solved by the Egyptians (see page 48). The salient difference was that their lands were drained by two main rivers, the Euphrates and the Tigris. Theirs was a clay soil which was flooded by heavy rainfall. The floods, unlike those of the Nile, were quite unpredictable. They were believed to be under the control of two gods, Nin-Gersu and Tiamat, both evilly disposed towards the human race. Because of the clay soil, drainage rather than irrigation was essential. The banks of the rivers were built up to prevent the spring flooding caused by melting snow in the mountains of Anatolia. Under good management, the land could produce two harvests every year. But to a considerable extent this depended on teamwork and cooperation by the farmers across the land.

The calendar played a key role in predicting and controlling all this activity. The phases of the Moon provided the means of counting off the seasons. The four phases of the Moon gave the lunar month, which could be either 29 or 30 days long. One or two feast days added to each month brought the official calendar in line with the solar year. This started with the spring equinox (when daylight lasted as long as darkness). As already indicated, the year consisted of 365 days, divided up into 12 lunar months of 30 days each, plus 5 extra days. To persuade ordinary people to accept this calendar, they were told that it had been ordained by Shamash, god of the Sun.

Rooting their calendar in astronomy gave the Babylonians a strong motive to work out new ways of dealing with number data. It was a two-way process. Their knowledge of number improved as they made more discoveries about the heavens; this enabled them to make more discoveries. For example, their number base of 60 gave them a better hold on fractions. Unlike Egyptian unit fractions (where the numerator, as we call it – the top figure in a modern fraction – was implicit, being equal to one), in Babylonian fractions the single number stood for the numerator; in this case the denominator was implicit as some power of 60, its value being given by its position. (Sexagesimal fractions were so useful that the Babylonian system was used in astronomy until the 15th century, by Kepler, Galileo and other worthies. It is still used to measure time.)

The Babylonians' intensive and rigorous study of the sky over many hundreds of years bears all the marks of a developed science. In addition to celestial observation, they carried out complex calculations concerning the duration of light and darkness in different seasons and in different places in the world as they knew it. From their observations of the rising and setting of the Moon and its movements, they worked out astronomical tables which enabled them to predict lunar eclipses. They were also interested in other heavenly bodies, especially the planet Venus, known as the 'evening star' (as when it appears before moonrise) and 'morning star' (as when it appears before sunrise).

All this activity had a darker side. The scientific impulse is to find connections between phenomena, and to develop a general explanation in terms of laws of nature which underlie these connections. The Babylonians, like many others, explained the underlying realities not by scientific fact but by inventing a pantheon of gods, existing in an invisible spiritual universe. It was the activities of these spiritual forces, they claimed, which lay behind the observed connections.

As a first step towards this kind of explanation, they believed that certain signs and omens predicted the future, without being too precise about the nature of the link between omen and outcome. Thus, they wrote down each sign and the kind of predictions that could be made from it. Two examples are, 'If the sky is dark on the first day, the year will be bad. But if it is clear when the new moon appears, then the year will be a happy one', and 'If on Sabatu 15th, Venus disappears in the West and remains away from the sky for three days, appearing then in the East, there will be catastrophes of kings'.

Babylonian omen literature is a vast collection of forebodings and signals of disaster – this is the dark side. The texts, written on baked clay tablets, go back as far as 1800 BC. They refer to the community rather than to individuals. Their primary purpose was to inform the king and his officials about impending military actions by hostile nations, and threats to the royal family. The official interpreter of omens also had the duty of giving advice on how to avoid the predicted misfortunes and to take advantage of the favourable omens.

These first attempts to read the future from portents such as monstrous births, weather signs and auguries took on more systematic form when the 'skill' of reading sheep's livers was invented. The astronomers identified star clusters as outlines of animals and humans. The Sumerians had observed and named three star clusters or constellations: the Bull (Taurus), the Scorpion (Scorpio) and the Lion (Leo). The Babylonians added nine others, and the twelve are now called the signs of the zodiac. The twelve constellations so identified occupy a relatively narrow band of the sky, in width about one-twentieth of the whole.

The daily eastward rotation of the Earth on its polar axis gives us the illusion that it is the sky which is going around. It seems to rotate westwards with uniform speed through a full circle. Just like the Egyptians with their decans (see page 51), so the Babylonians thought that the 12 constellations they named as animals, or 'signs' of the zodiac, were way-stations, hostelries where the Sun, Moon or a given planet tarried a little before continuing across the sky. The coincidence of each celestial body with each star sign was an omen of good or bad fortune, depending on the totality of all indicators. There was no suggestion at this time that such conjunctions rigidly decided the future. They were mere pointers to what could happen if no precautions, or helping actions, were taken by individual or state.

Omens of good and bad fortune were classified according to whether they were associated with the Moon goddess (Sin), the Sun god (Shamash), the weather god (Adad) or the planet Venus (Ishtar). Thus, the variety of indications – lunar and solar eclipses, haloes, crescents, two Suns in the sky, thunder and lightning, cloud formations, earthquakes, planets seeming to stand still, conjunctions between the rising or setting of the Sun or Moon and a given planet – testifies to the fact that no sort of general theory of causal relation had yet been worked out. Astrology was struggling to be born. It was (some) Greek philosophers, centuries later, who developed Babylonian star-charts into a full-blown system of prediction of individual destiny and character from the position of the celestial bodies at the moment of birth.

1 2 3 4 5 6 7 8 9 10 11 12 13 14 15 16 17 18

ANCIENT EGYPT

Concerning Egypt, I shall extend my remarks to a great length, because there is no country that possesses so many wonders.

Herodotus (5th century BC)

I had seen a hundred things, while a thousand others had escaped me; and, for the first time, had found access to the archives of the arts and sciences.
Baron Vivant Denon (Napoleon's art adviser in the 1803 Egyptian Campaign)

As we saw in Chapter 1, the first step towards written numbers was taken when tally marks came into use, probably in pastoral societies, to record the counting of relatively large numbers of animals. In ancient Egypt, possibly as much as 4000–5000 years ago, the priests and scribes took a step further by inventing a system of numerals which varied according to the size of the number. To report a total, you gave the individual numerals, and the number of each in the grand total. Using these number-signs, the Egyptians could add, subtract, multiply and divide. But they had no special symbols for these operations. Instead they used the system now called 'rhetorical algebra': alongside the numeral, they gave a form of words describing what had to be done.

Apart from religious observance, one of the main duties of Egyptian priests was to record such things as wars, royal decrees and the history of each reign. To do this, they used sacred writing (hieroglyphics). Hieroglyphics were reserved for formal, official inscriptions, and are the picture-writing we see in royal tombs and temple walls, painted on or incised in stone. Numbers were seldom used in hieroglyphic writing. This is because they were rarely needed in formal inscriptions. In recording details of a conquest, for example, verbal descriptions ('all-conquering armies'; 'countless captives') gave the pharaoh's achievements more prestige.

Hieroglyphics were too complex for ordinary purposes, so the scribes used a cursive version, a kind of shorthand written in ink with a stylus, on papyrus. Such writing is known as hieratic ('temple writing'). Its signs for numerals are shown on page 42.

Scribes were of a different class from priests. They were often trusted slaves; they might be freed for diligent service. They were engaged as secretaries and accountants; they worked in temples, which also served as government offices. They wrote letters to dictation and kept official records. They probably worked out the calculations for such things as government projects in rough, doing their figuring and verifications on scraps of papyrus which were thrown away as soon as used.

Unfortunately, all that we know about Egyptian mathematics is preserved in, and has to be deduced from, no more than two papyrus rolls, a few fragments of papyrus and a scrap of incised leather. The most important of these relics is the Rhind mathematical papyrus, now in the British Museum (there is a second copy in Moscow). It was bought by the Scottish antiquary Alexander Rhind, while on holiday in Egypt in 1858. It was compiled in the 16th century BC by a scribe called Ahmes (or Amos; the hieroglyphics are the same), who is our principal source of information about the intricacies of Egyptian mathematics.

Egyptian numbers

Egyptian numbers, like Egyptian words, were written from right to left. After the fashion of the time, Ahmes did not use – or have – any signs for equals, for plus or minus, for multiply or divide. He wrote fractions as single numbers with a dot over the top (e.g. $\dot{5}$): denominators without numerators. (They are called 'unit fractions', because the numerator is always the same: one. In modern equivalents, $\dot{5}$ would mean $\frac{1}{5}$). This use of a dot to indicate a fraction is the hieratic equivalent of the hieroglyph for an open mouth ($\wp$). It suggests that the original use of fractions was to divide out shares of food and drink – and indeed many of Ahmes' problems in the Rhind papyrus are about dividing up loaves of bread and jugs of beer. Our modern division sign (a bar) may have had a similar origin.

The layout of each group of numbers implicitly states the nature of the problem to be solved. There are also half a dozen set phrases, one of which is used to pose the question. For example, Ahmes states a rule for finding two-thirds of a fraction (a basic rule in Egyptian arithmetic) as follows:

The making of two-thirds of an odd fraction: If it is said to thee, 'What is two-thirds of a fifth?', make thou twice of it, then six of it.

This is one of the earliest known algorithms in the history of mathematics. In contemporary terms, Ahmes is saying that two-thirds of one-fifth is equal to half of one-fifth plus one-sixth of one-fifth, that is one-tenth plus one-thirtieth. To spell out his algorithm: multiply the original fraction by two and write down the result. Then multiply the original fraction again by six and add the two results. In modern notation,

$$\tfrac{2}{3} \times \tfrac{1}{5} = \tfrac{1}{10} + \tfrac{1}{30} = \tfrac{4}{30} = \tfrac{2}{15}.$$

Egyptian arithmetic has long been devalued because it lacks a sign for zero and has no place-system. One might as well criticise an Eskimo for not wearing top hat and tails when he goes fishing. Egyptian scribes used a completely different set of rules. They indicated units, tens and hundreds by means of different numerals, not by position. If you have different signs for units, tens, . . . and so on, it is irrelevant in what order you write them down. One of the advances Egyptian mathematicians made on the Babylonians was to indicate which parts were fractions and which were whole numbers. The notion that different symbols should be used for different 'levels' of tens made the zero unnecessary.

This system was simpler, and far less tedious, than ours. Some of the dull things European children have for centuries had to learn in school, the Egyptians made no use of at all: percentages, money conversions, the lowest common denominator, and so on. The ancient Egyptians were spared the Industrial Revolution; their children were spared commercial arithmetic. Egyptian arithmetic was certainly easier and, as far as fractions go, more precise than ours. For instance, there were no recurring decimals (which stand for indeterminate numbers), neither did they round off figures when dividing.

Egyptian methods of teaching arithmetic were advanced, especially by later European standards. Plato, who spent some years in Egypt as a student, is our authority for this. In his dialogue *Laws* (freely translated into modern idiom by Jowett), he writes about the Egyptians:

they teach arithmetic to children at the same age as they are learning to read and write. Their teaching takes the form of pleasant games like dividing out apples and flowers, now to a large group, now to a smaller one. Again, they take vessels of gold, silver and brass, mix them up and then sort them out. They freely adapt the games to the numbers available. In this way, they enable the pupils to know about such things as the movements of armies and supplies. They learn how to manage a household. The pupils are more alert and in touch with reality. They learn how to measure and count. In this way they are better able to deal with things around them.

These Egyptian ideas resurfaced in 20th-century British schools, as 'modern maths' (*sic*), but failed to catch on, largely because of lack of analysis related to the age of the class involved.

To complete the picture, it has to be mentioned that, when the Rhind papyrus was composed, the Egyptians had no metal currency. The importance of a monetary system in the development of arithmetic (especially fractions) is difficult to exaggerate. Loaves of bread and jugs of beer can be divided physically into any size, down to individual crumbs and drops. But this is impossible with metal coins. In fact, lack of a coinage may account for the Egyptians' unusual approach to fractions and the emphasis on bread and beer in the surviving accounts of their arithmetic.

Using the system described, the Egyptians could carry out the most extraordinary and complex calculations. Their elaborate funerals for royal families and top bureaucrats involved much detailed figuring. Running an empire and waging wars of conquest demanded logistics for the victualling and transport of large numbers of soldiers and masses of materials. Sophisticated calculations were needed to plan and build cities and monumental buildings, which are still a wonder in the modern world. They could balance their accounts. They could check on contractors. They could look out for swindles. They could record the number of captives available as slave labour and share them out for public works. They could estimate how much food and drink, how many blocks of stone of different shapes and sizes, how many slaves and overseers would be needed from day to day to build the pyramids. They could reckon the dates of completionof the various stages of the work, using the most rational calendar ever invented (superior to the version we use). They could reckon how much grain was needed to make loaves of different nutritional value, or beer of various strengths. Calculation was essential to the running of the state as a tight, efficient system.

1	10	100	1,000	10,000	100,000	1,000,000

Egyptian numbers started at one and went up as far as a million. One was symbolised by a papyrus leaf; 10 was a tie made by bending a leaf; 100 was what looks like a piece of rope; 1000 was a lotus flower; 10,000 was a snake; 100,000 was a tadpole and 1,000,000 was a scribe raising both arms above his head as if in astonishment. Using these symbols, an ancient Egyptian scribe time-transported to New York City in 1975 might have recorded its population (9,526,863) as 9 astonished scribes, 5 tadpoles, 2 snakes, 6 lotus flowers, 8 pieces of rope, 6 ties and 3 papyrus leaves:

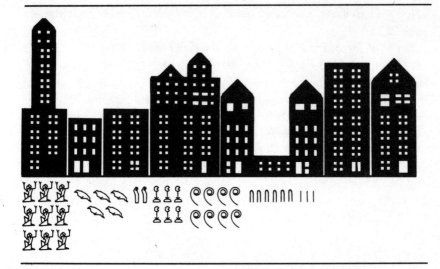

In hieroglyphics, reading the complete number is about as easy as in the tally system. You count up each kind of numeral, as on your fingers. If you get to nine, you have exhausted the particular species of numeral you are counting; you go on to the next highest symbol, and continue counting. Subtraction is the reverse of addition, and is equally straightforward. You 'take away' the numbers of each species indicated. You need only to remember the rule about borrowing one (and paying it back) from the higher-level number when you run out of the number below in the system.

Tables

Tables (calculation charts) were invented in Babylon, but Egyptian mathematicians developed and perfected them in forms that were used, unchanged, for millennia. The Rhind papyrus gives clear evidence of a table for addition, which could also be used for its complement, subtraction. There was also a table of unit fractions, so that long strings of fractions could be added routinely. (It is said that in the Rhind papyrus a string of 16 fractions is added, correctly, but I have so far not managed to locate this.)

Multiplication

The Egyptians had no need to learn the 'times tables' for multiplication or division. The insight they had was that any whole number could be made up by adding selected terms from the binary series: 1, 2, 4, 8, 16, 32, 64, 128,

256, . . . (This idea is central to the workings of the modern computer; see page 232.)

Suppose, for example, that an Egyptian scribe wanted to multiply 256 by 17. To get 17 we add 1 and 16 from the binary series. As the series shows, we reach 16 by doubling 1, then doubling the answer, then doubling again, then redoubling. So, to calculate 17 times 256, we simply redouble 256 four times and then add the result to the original number 256. Using modern numerals, the Egyptian scribe might write this in a table as follows:

$$
\begin{array}{ll}
1^* & 256^* \\
2 & 512 \\
4 & 1024 \\
8 & 2048 \\
16^* & 4096^* \\
17 & 4325 \\
\end{array}
$$

(*add these numbers together*)

Or, set out completely in modern notation:

$$
\begin{aligned}
1 \times 256 + &= 256 + \\
2 \times 256 &= 512 \\
4 \times 256 &= 1024 \\
8 \times 256 &= 2048 \\
16 \times 256 + &= 4096 + \\
\end{aligned}
$$

(*add the numbers with plus signs*)

$$
\begin{aligned}
17 &= 1 + 16 \\
17 \times 256 &= (1 \times 256) + (16 \times 256) \\
&= 256 + 4096 \\
&= 4352 \\
\end{aligned}
$$

This method works just as well, and just as simply, in more complex calculations:

To multiply 226 by 13
$$
\begin{aligned}
1 \times 226 + &= 226 + \\
2 \times 226 &= 452 \\
4 \times 226 + &= 904 + \\
8 \times 226 + &= 1808 + \\
\end{aligned}
$$

(*add the numbers with plus signs*)

$$13 = 8 + 4 + 1$$
$$226 \times 13 = (226 \times 8) + (226 \times 4) + (226 \times 1)$$
$$= 1808 + 904 + 226$$
$$= 2938$$

This looks cumbersome to set out, but Egyptian scribes, used to such calculations, could presumably perform them very quickly, with minimal need to write things down. (They would, for example, have been familiar with the binary series (see page 43), which is the key to the whole of Egyptian arithmetic.) The method works just as well with fractions.

Division, including fractions

Perhaps the most dazzling mathematical insight of the Egyptians was that the four arithmetical processes are closely related. Multiplication and division, like addition and subtraction, are mirror images of one another. (The calculation on page 44, for example, tells us not only that 256 multiplied by 17 equals 4352, but also that 4352 divided by 17 equals 256.) The basic process, fundamental to all the others, is addition. This insight is shown in many places in the Rhind papyrus – and modern computers also make use of it (see page 232).

Using the binary series, exactly as for multiplication, an Egyptian scribe seeking to divide 256 by 17 would probably have consulted a division table, and then written:

$$
\begin{array}{ccc}
17 & 256 & \\
1 & 15 & \overline{17}
\end{array}
$$

The modern equivalent would be:

$$1 \times 17 = 17$$
$$2 \times 17 = 34$$
$$4 \times 17 = 68$$
$$8 \times 17 = 136$$

$$256 = 136 + 68 + 34 + 17 + 1$$
$$\tfrac{256}{17} = 8 + 4 + 2 + 1 + \tfrac{1}{17}$$
$$= 15\tfrac{1}{17}$$

Once again, the method also works for fractions. In passing, it should be said that the novelty of Egyptian algorithms, coupled with our lack of familiarity with their notation for fractions, makes Egyptian work with fractions quite difficult to understand. As moderns, we have learned different drills, to solve fractions in modern notation. To an Egyptian, versed in the drills and notation of his or her own time, fractions would have posed few problems.

In real life, of course, not all fractions are unit fractions. For example, $\frac{2}{7}$ or $\frac{43}{180}$ could quite easily be the solution to a problem, but it is not at all easy to see how to write them in unit fractions. Each such fraction would be written as a string of unit fractions, which would add up to give the $\frac{2}{7}$ or $\frac{43}{180}$ or whatever. If we use the word 'expression' for such a string of unit fractions, and taking our cue from the fractional answers in the Rhind papyrus, we can work out the rules for deciding what the best such expression would be.

The first rule is that all fractions are unit fractions (such as $\frac{1}{2}$; $\frac{1}{3}$; $\frac{1}{15}$ and so on), except that the Egyptian scribes did not need to use any numerator (the number on top of a fraction). The second rule is that we try to find expressions always with even numbers as denominators: $\overline{4}$, $\overline{68}$, $\overline{92}$ etc. The third rule is that no fraction is repeated. The fourth and most binding rule is that we keep all fractions in the expression as small as possible. ('Small' in this context means 'with a small denominator': $\frac{1}{2}$ is a 'small' fraction; $\frac{1}{2378}$ is a 'big' fraction.) The fifth and last rule is that our expression must contain as few unit fractions as possible, never more than four. For example,

$\frac{2}{27}$ is equal to $\overline{18}$ $\overline{54}$
$\frac{2}{37}$ is equal to $\overline{24}$ $\overline{111}$ $\overline{296}$
$\frac{2}{47}$ is equal to $\overline{60}$ $\overline{332}$ $\overline{415}$ $\overline{498}$

The mystery of the Rhind papyrus is the question: How did the priests discover which expressions best satisfy these five rules? The values given in the table of expressions for fractions, which make up a large part of the document, are close to the 50 best out of 22,295 combinations generated by computer, and checked, by the present author. The mystery disappears as soon as we do the divisions exactly in the way the scribe himself would have done them, and forget about modern arithmetic.

A second question now arises: Since the rules seem arbitrary, how did they arise? It seems unlikely that they reflect the way the answer is implemented – for example (reverting to the bread and beer used in the Rhind papyrus examples), by the cutting up of actual bread and dividing out of actual beer. It seems incredible that (say) to divide two loaves between 83 men ($\frac{2}{83} = \overline{60}$ $\overline{356}$ $\overline{534}$ $\overline{890}$), the scribe would have contemplated dividing two loaves into 60, 356, 534 and 890 pieces, rather than the much simpler division of each loaf into 83 pieces. The best explanation seems to be that the purpose of these examples had nothing to do with dividing actual bread and beer: it was to illustrate how to work with fractions. The table, in short, may have served a purely didactic purpose.

It is very difficult for any specific set of unit fractions to satisfy all five rules at once. For example, the expression given on page 46 for $\frac{2}{37}$ ($\overline{24}$ $\overline{111}$ $\overline{296}$) fails to keep the second rule given. The scribe would have to make a subjective decision about the best choice of fractions to represent the number. The complex nature of the rules makes the feat of balancing them, one against the other, very difficult. This is, presumably, why the table was made up. All the difficult calculations and decisions had already been made in advance, and the solutions were ready to use. The answers to problems could be found mechanically and therefore quickly, in accordance with the rules.

The tables (presumably official documents on papyrus) would have an aura and would give the scribe authority. If people felt like arguing about their shares, the tables would serve to silence them. A modern analogy is our income-tax tables, which are ostensibly designed to save bother, but in fact inhibit most of us from arguing with the tax official. (The final remark about Egyptian fractions is that the priests used a formalised drawing or painting of the 'eye of Osiris' to locate (and illustrate) various fractions such as $\frac{1}{2}, \frac{1}{3}, \frac{1}{5}$ and so on. This is merely another complication. It is not dealt with in the Rhind papyrus.)

The Rhind papyrus is the earliest textbook on arithmetic we have – and it has all the defects we associate with didactic mathematics. It is as elusive and perfunctory as the average modern computer manual – well, almost. One-sixth of it seems to consist of exercises illustrating the use of the table of fractions. Perhaps the whole text is no more than exercises, with examples from real life to enliven them.

As against this view, we can cite a second document, the records of the Temple at Illahun, as translated by Borchardt. It refers directly to the payment in kind (bread and beer) of all the staff who worked at the temple. In a slightly adapted form, this document is set out in the table overleaf.

There is, of course, a simpler and more practical way to solve problems of dividing bread and beer. It did not take army cooks several millennia to discover that 23 loaves could be divided equitably between 17 men by assigning each man one loaf and dividing the other six into thirds. This meant a loaf and a third for each man and a third left over for the cook. (Of course, this is not how it would be reported to the officer of the day, but that is another matter.)

On the other hand, the Egyptian priest–teacher seemed to be concerned with the most precise solution. Learning how to divide bread and beer as accurately as possible, as an academic exercise, was a useful preparation for

Personnel	Number of portions	Loaves of bread	Jugs of bitter beer	Jugs of mild beer
Director	10	16 $\overline{3}$ (ie 16$\frac{2}{3}$)	8 $\overline{3}$ (ie 8$\frac{1}{3}$)	27 $\overline{2}$
Head priest	3	5	2 $\overline{2}$	8 $\overline{4}$
Head reader	6	10	5	16 $\overline{2}$
Scribes	1 $\overline{3}$	2 $\overline{6}$ $\overline{18}$ (ie 2+$\frac{1}{6}$+$\frac{1}{18}$=2$\frac{2}{9}$)	1 $\overline{9}$	3 $\overline{3}$
Reader	4	6 $\overline{3}$	3 $\overline{3}$	11
Priest 1	2	3 $\overline{3}$	1 $\overline{3}$	5 $\overline{2}$
Priest 2 (3)	6	10	5	16 $\overline{2}$
Priest 3 (2)	4	6 $\overline{3}$	3 $\overline{3}$	11
Cleaner	1	1 $\overline{3}$	$\overline{3}$ $\overline{6}$	2 $\overline{2}$ $\overline{4}$
Guards (4)	1 $\overline{3}$	2 $\overline{6}$ $\overline{18}$	1 $\overline{9}$	3 $\overline{3}$
Watchmen (2)	$\overline{3}$	1 $\overline{9}$	$\overline{2}$ $\overline{18}$	1 $\overline{3}$ $\overline{6}$
Workers (2)	$\overline{3}$	1 $\overline{9}$	$\overline{2}$ $\overline{18}$	$\overline{3}$ $\overline{2}$
Totals	42	70	35	115 $\overline{2}$
Each portion	1	1 $\overline{3}$	$\overline{3}$ $\overline{6}$	2 $\overline{2}$ $\overline{4}$

Notes: $\overline{x}$ stands for the unit fraction $\frac{1}{x}$: for example, $\overline{6}$ is $\frac{1}{6}$.
$\overline{3}$ stands for $\frac{2}{3}$.
It will be seen that the priest's calculations – or those of his amanuensis – are not infallible. In some places, $_{\overline{3}}$ is written instead of $\overline{3}$, and the 'totals' line is open to argument.

more important calculations. Knowing the algorithms, students could be prepared for really vital problems, such as: how do you place a narrow opening so that twice a year forever (on our 20 October and 20 February) the Sun will shine on the face of Rameses II in the royal tomb at Abu Simbel? This brings us to the most important function of all 'learned' mathematicians in ancient Egypt: calculating and organising the calendar.

The Egyptian calendar

The one crucial element in the survival of the Egyptian people was the success of peasants in providing an agricultural surplus for the towns and for trade. In turn, this was linked to the ability to predict the beginning and the duration of each of the three seasons recognised by the Egyptian farmer: the annual flooding of the Nile delta, the period of seed-time and growth and the period of harvest.

The moment in time when this sequence began each year was easily established, and could be used as a fixed point every year. But the length of time between floodings, and consequently the length of each year, was irregular: between 1945 BC and 1875 BC, for example, the year varied between 345 days and 415 days, a difference of over two months.

However, for the purposes of agriculture, a reasonable variation in the date of onset of the seasons was not crucial. The harvest also depended on other factors, many of them – crop disease, insect infestation, seed quality – even less predictable. It was more important that, in spite of large variations, the records kept by the priests showed that the average interval between floodings worked out at 365 days, close to the astronomical year, which we now know to be 365.4511 days. (The discrepancy, arising from the Egyptian method of averaging, which ignored fractions, had important implications for the history of the calendar. Julius Caesar adopted the Egyptian calendar, with minor alterations, for the Roman Empire in 45 BC. This 'Julian' calendar, stripped of its cumbersome system of numbering days in each month according to their position before three fixed points, the Kalends, Nones or Ides, was later passed on by the Christian church to other countries which came under its influence.)

The number of days in the Nilotic year (365) was the integral value (i.e. without fractions of days) of the solar year. Priests soon noticed that the start of the flooding was heralded by a celestial messenger, the dog-star Sirius, known to the Egyptians as Sothis. They made this link so often that they came to believe that the star was the cause of the annual flooding. Since Sothis is the brightest star in the sky, it is not easily overlooked. They observed that it appeared just before sunrise each year, after an interval of $365\frac{1}{4}$ days. This appearance of the star shortly before the Sun rises above the horizon, its 'heliacal rising', happens only once in 365 days, the time it takes for the Earth to complete its journey round the Sun. The Egyptians thus adopted the solar year (365 days and nights, the interval between successive heliacal risings of Sothis) as the basis of their calendar.

Pyramid texts of the 5th dynasty (2,400 BC) show that the Nilotic year and the Sothis calendar were already established by this date. The 5th dynasty was believed to have its origin from the union between the Sun god, Ra, and the wife of one of his priests. Thus the link, however mythical, was made between the Sun, pharaoh and the Nile. Priests and the ruling dynasty alike made a connection between heavenly events and the main agricultural happening of the year.

Apart from the role of Sothis in marking the beginning of the year, the heavens played a further part in enabling the priests to work out details of the solar calendar. The year was divided into three seasons, each of four months, the twelve months so formed being standard at 30 days each. This left five days over. These five days were special, known as 'temple days' or 'epagonal days' (that is, set aside for festivals). The quarter of a day left over in the solar cycle remained a source of error. It meant that the solar calendar (365 days) was ahead of the solar cycle (the Earth going round the Sun in $365\frac{1}{4}$ days), and that the discrepancy increased with every passing year. In four years, the difference would be a complete day. In a century, the calendar might say it was winter while the Sun was blazing overhead.

The discrepancy led to continued use of a lunar calendar as well as of the civil calendar of 365 days. The lunar calendar was used to date religious festivals. Its basis was the different phases of the Moon. (The Moon goes through a cycle of four changes as it waxes and wanes. In the first phase, the image we see is a right-facing, thin crescent which gradually extends to a half-moon. In the second phase it expands further to become a full circle (full moon). In the third phase it wanes (lessens) until it becomes a half-moon. In the fourth phase it changes to a left-facing crescent, which then thins out until the Moon disappears altogether. This sequence from start to finish, the 'period' of the Moon, takes between 29 and 30 days, and each of the four phases takes just over a week.)

Each of the 12 months of the lunar year was $29\frac{1}{2}$ days long. Since Egyptian priests reckoned averages as whole numbers, without using fractions, they calculated some months at 29 days, others at 30 days. This made the lunar year of 354 days shorter than the solar year by 11 days, a much more serious discrepancy than the quarter-day of the solar year. It meant that, very quickly, calendar and seasons would be quite discordant. To restore harmony between the calendar and the natural round, and between the gods and the calendar days, the eleven missing days had to be produced from somewhere. An extra month was inserted here and there in the lunar calendar, as necessary. Sometimes a pious fraud was used: a particular month would be run through twice, as though by mistake.

The Egyptians invented a third calendar by a procedure that led to our present-day division of the day into 24 hours. This was a much more complicated calendar since, in addition to the Sun and Moon, it required observation of 36 different star constellations. It worked as follows. Egyptian star-watchers continued to track Sothis through the heavens, for a period of 10 days after its first appearance. Each position occupied by the star on these days was noted and became a marker for one of the hours of darkness. On the 11th day another star or constellation was chosen as marker, on the basis that it was now the heavenly body whose rising immediately preceded sunrise. This star or constellation was observed for the next 10 days, each position in the heavens being taken as marking the hours after ten.

In all, 36 heavenly bodies were chosen in this way, their positions in the night sky being noted for 10 days as they ran their courses. The 36 heavenly bodies were known as the 'decans', a word referring to the 10 days assigned to each star and to the 10 places in the sky that it occupied during these days. Thus, 360 days were assigned, 10 to each of the 36 decans. Perhaps of more significance, the decans also performed the function of indicating the time it took for the star to pass through one position in the sky and travel to the next. Since stars appear to the observer to be moving at constant speed, these time intervals were taken to be equal. The average time taken by the 36 decans to traverse the night sky was found to be 12, the same average time as the Sun takes to traverse the sky during the day. Thus 12 'hours' were calculated to be the average length of darkness as well as of day. This gave a total of 24 hours for the duration of night and day.

This Egyptian system of observation was very like the division of the sky into 'lunar mansions' by Indian astronomers, or the 'houses of the zodiac' devised by Babylonian astronomers. But the Egyptians turned the system to practical advantage as a measurement of time, devising equal units. The decan calendar seems to have had some significance to the afterlife: what seem to be star-decans were often painted on the lids of coffins.

It is a mistake to imagine that the four different calendars were in any way in competition with each other, or that they were four attempts to devise a 'true' calendar, one that would reflect without error the real state of affairs with regard to the heavenly bodies. Such notions of singularity came much later and were linked to the scientists' concern with developing a system of measurement based on as few concepts as possible. The Egyptians used each of their four systems in a different situation and for different purposes. We might find an analogy in the way people used different weights, until recently, for different purposes. It was only at a relatively late date (just after the 1789 French Revolution) that the same system was introduced in France to weigh (say) gold dust, grain, fruit and fish. In fact, there is still no standard way of selling certain fruit as units, or by weight. Oranges are sold mostly by the unit; pears indifferently by both systems. No confusion is created in the minds of shopkeepers or customers as they move from one item to the next.

1 2 3 4 5 6 7 8 9 10 11 12 13 14 15 16 17 18

Ancient China

Practice without theory is blind, theory without practice is sterile.
Immanuel Kant, as amended by Karl Marx

In our time, China has yielded its previously hidden intellectual secrets to Western scholars. Managing to overcome endemic Western chauvinism and the one-eyed perspectives of their disciplines, these modern scholars have focused on the total picture of Chinese knowledge. Rather haltingly as yet, we in the West can begin to see ancient Chinese civilisation as a unity, as the Chinese themselves do.

Nowhere is this unity clearer than in the history of calculation. Europe struggled for centuries with the legacy of Greece and Rome, selectively transmitted by the Roman Catholic church. Until about 1550, the dead hand of Greek and Roman number systems hobbled merchants and accountants in their work. Greek ideas – that mathematics was geometry, that arithmetic was the theory of divine numbers and that science was abstract reasoning – crippled Western science and mathematics for centuries, until the Western scientific renaissance gave rise to major breakthroughs by Galileo and Newton in the 16th and 17th centuries.

For centuries before the West, China had a tradition of scientific knowledge as the integration of practical experience and theoretical analysis. Chinese scholars were trained in analytical, abstract thinking. Some four centuries before Christ, they perfected a decimal number system and the calculation methods that could be used with it. They discovered the essential features of numbers some 2000 years before Europe. They could solve simple, simultaneous, quadratic and other equations of any degree up to ten. They invented, and made systematic use of, such valuable aids to calculation as the abacus and the counting board. While most European scholars were maundering on about witches, or wasting time and energy in royal or ducal courts as upper-class domestics, Chinese scholars were busy about the practical affairs of state.

This excellence had three main causes. First, and most important, was the Chinese view that calculation, so far from being a low-level skill suited to slaves and domestics, should be the concern of the finest, most highly-trained intellects in the state. Abstract thinking was a way to solve the

53

problems of everyday living – and those problems were the concern of all, but especially of state officials, appointed by the emperor. Correct calculation was essential for the conduct of state affairs. The emperor, by being just and fair, upheld order in the universe. To advise him how to keep a proper balance, his officials had to observe and calculate the balance of forces in the universe. These things required wisdom and scientific knowledge, and – in a typical piece of logical thinking – the Chinese encouraged excellence in these fields by a system of competitive examinations, opening the hierarchy of government to any (male) person of sufficient intellectual calibre.

The second special advantage the Chinese had over most other ancient nations was the elegance and simplicity of the Chinese language. The spoken language consisted (and still consists) of one-syllable words. Complex ideas are expressed by word combinations, but each word retains its unity – the language does not run words together, as do (for example) Eskimo, Finnish or German. There are no past, present or future tenses. There are no gender, mood, tense, case or other variables. There are no definite or indefinite articles. What we call adjectives do not need to agree with what we call nouns. There is no essential word order: the way you speak is determined by what you want to say, and you add force and expression not by word-changes but by pitch and tone of voice.

The third crucial factor in the development of Chinese mathematics was the nature of Chinese writing. Written Chinese is a picture language, in much the same way as Egyptian hieroglyphics. Each symbol originated in the representation of an object or idea. Symbols are combined to represent complex ideas. This fact – so different from, say, European phonetic alphabets (where symbols represent sounds) – made it possible for the language to be understood all over China by literate persons, no matter what local variants existed in the spoken language. All educated Chinese shared a common language – and it was a language of ideas.

The robustness and stability of Chinese mathematics are shown by the fact that their system remained in use, virtually unchanged, for two millennia, until it began to be replaced in the present century by the Arabic number system used throughout the world. The most comprehensive description of ancient Chinese mathematics was written, not in ancient times but in the 13th century, by the scholar Chin Chiu-Shao (1202–61). He studied astronomy and mathematics, worked as a local official in Szechuan and Wu-hsing, and in 1247 wrote *Shu Shu Chin Chiang* ('Mathematical Treatise in Nine Sections'). Each section of this book contains two chapters, and each chapter deals with nine problems and how to solve them. The problems are practical – the sections deal respectively with barter and purchase, military arithmetic, building work, money and grain, taxation, surveying, land-measurement, astronomical calculations and indeterminate equations – and the book gives a clear idea not only of the scope and skills of Chinese mathematics, but of its intimate interrelationship with the practicalities of everyday administration.

Chin lived at a time of great political instability. The Mongols, led successively by Kublai Khan and Genghis Khan, were riding throughout Asia, conquering and pillaging. Chin himself fought them for ten years on the frontier, and complains that this distracted him from publishing his book. At the same time, the 13th century was the golden era of Chinese mathematics. Chin says that there were more than thirty schools of mathematics in China, and three other major scholars are known to us from this time, which was particularly famous for its algebra. This is a level of activity, and of achievement, to rival any other period of mathematical research, anywhere in the world, until modern times.

Chinese numerals

In ancient China, five types of written numerals existed, each used for a different purpose. The two most important kinds were stick numerals (derived from the wooden sticks that represented numbers on a counting board: see page 56), and the basic numerals. The other three kinds were, more or less, elaborate derivatives – for example, the ornate 'official numbers' used on banknotes, contracts and business documents to deter forgery. (There were many different, idiosyncratic forms, some of which are shown in the table below.)

Arabic numeral	Chinese name	Stick numeral	Basic numeral	Commercial numeral
0	ling	□	o	o
1	i	I	—	I
2	erh	II	=	II
3	san	III	≡	III
4	szu	IIII	四	✕
5	wu	IIIII	五	⅄
6	liu	⊤	十	⊥
7	ch'i	⊤⊤	ヒ	⊥
8	pa	⊤⊤⊤	ハ	≝
9	chiu	⊤⊤⊤⊤	升	夊
10	shi	—	十	十

Spoken numbers greater than ten (*shi*) were formed by combining ten with the other numeral, as in *shi-san* (thirteen) or *san-shi* (thirty); 100 was *pai*, 1000 *ch'ien*, 10,000 *wan*, 100,000 *pai ch'ien*, 1,000,000 *wan wan*. The idea of a large number of things was also expressed by using the numbers eight or ten in combination with the name of the thing or idea. Still larger numbers were indicated in the same way, using 100 or 1000 – and these combinations sometimes led to spectacularly poetic descriptions of reality:

100 thing	= everything
100 times think	= prime minister
100 worker	= the working class
1000 direction	= versatile
1000 00 month	= great antiquity

Central to Chinese mathematics, and to the running of the Empire with which it was so intimately involved, was the calendar. The calendar dictated the timing of Imperial rituals in the Forbidden City, and was changed whenever a new Emperor was installed. The adoption of each new calendar became a test of loyalty in all regions of the Empire. Concern with agriculture as the basis of the economy set problems related to volumes of water for irrigation, land surveys and the division of land. Warfare, and the assault on walled cities; divination based on magic squares and numerology; taxation of land and crops, using clandestine means of calculating taxes on income to uncover fraud; calculation of the size of populations and working out the results of the census; money lent at interest with repayment schedules – all these gave rise to the problems solved by Chinese mathematicians from earliest times.

The counting board

Chinese calculation was done mentally; devices like the counting board were used, not to do calculations but to keep track of them. The counting board was made of wood and marked off in squares, like a large chessboard. The person doing the sum laid out counting sticks on the board, to the required numbers and in the correct squares. Each stick was about four inches long, like a long matchstick. A full set contained 271. They were in two colours: red for positive numbers, black for negative. Each stick counted as 1, 10, 100 etc., according to the square it was placed in, going from right to left.

Having set up the sticks, the expert could proceed with calculation. The sticks were moved with lightning speed: it was as though the scholar's hands were swallows flying through the air. Using them to 'store' intermediate stages, and with the mental agility built up from years of training and practice, Chinese mathematicians were able to carry out the most extraordinary calculations with ease.

The use of sticks, laid on squares, is clearly the origin of the place system, and may also explain other features of the decimal system invented or formalised by Chinese mathematicians. The fact that some squares would be empty (have no stick on them) because the place in question was void explains how the zero came into existence: the rectangular (later round) zero represents an empty square. The three elements of Chinese number – the place of the numeral indicating its value; ten numerals being enough to represent any number however large, from zero to a thousand million less one; the empty place shown as a zero, are vital to the Chinese pre-eminence in ancient mathematics; all arise not from theory but from the simple, practical positioning of sticks on squares.

On paper, numbers were written in the form of stick numerals (see page 55). Each stick counted as one unit, up to five. Then, still in the same square, the next stick was turned at right angles to stand for five, with other 'unit' sticks added, parallel to the earlier sticks representing number 5, to indicate 7, 8 and 9.

The diagram below shows the counting board set up to assist calculation of a simple subtraction sum: 1,470,654 − 320,430.

Counting board at beginning of calculation   *Arabic equivalent*

I	☰	⊤⊤		⊤	☰	IIII	black sticks 1 470 654
	☰	II		IIII	☰		red sticks 320 430

Counting board at end of calculation   *Arabic equivalent*

I	—	IIIII		II	=	IIII	black sticks = 1 150 224

The 'sign', plus or minus, was given by the colour of the sticks: red for plus, black for minus. There were no signs for operations in Chinese mathematics; plus or minus signs in the tables in this chapter have been supplied.

To show the process in the diagram, we have left the sticks in place. In real life, the mathematician would subtract by clearing off sticks, square by square, from the second row, removing the same number of sticks in the corresponding squares in the first row. Addition would be done by placing the appropriate number of sticks in place, on the board. (There would be no third, 'answer' row on the real-life board. When all sticks had been placed or removed, the answer would remain.)

To do the sum, therefore, one begins by leaving the digit which stands for one million untouched. In the next column, three sticks are removed from each square, leaving one in the answer. In the ten thousands column, the seven sticks have two sticks removed at the same time as two sticks below are removed. In writing on paper, we would show a zero in the next column (the thousands place). The board would merely have an empty square, because there is nothing to borrow from it or carry to it. The six sticks in the hundreds place are reduced to two, and the four sticks below are also removed. In the tens place three are taken from above and three from below, leaving two in place on the board square. The units place remains as it is, there being nothing to take away. The subtraction is now completed, literally: surplus sticks have been taken from every line. Only one set of numbers remains, in the top line. It is the answer: 1,150,224.

As a result of training and practice, Chinese mathematicians would, in time, develop enormous skill and visual memory in using the sticks and board. In doing the subtraction above, for example, the mathematician would probably just glance at the table, sweep it clean and lay out the answer. To a mathematically sophisticated observer, the process looks like child's play – and indeed in the modern Cuisenaire method, based on Chinese practice, children learn arithmetic as a game with sticks. To the naive, non-literate observer, however, it must have seemed like magic, a casting of spells.

Two equations, two unknowns

Using the same kind of practical method, the Chinese discovered how to solve simultaneous equations: those in which two quantities are unknown. For example, to find out the numbers of male guests and female entertainers at a conference at a tea house in Soochow, one needs to know two number facts about them, each giving us some number relation between the two unknowns. In this case, let us assume that the guests and entertainers used 52 dishes at dinner, and that they played a group game afterwards, when the men and women were paired off.

In modern style, using symbols to 'stand in' for people, we can write m to mean the number of men, w the number of women. The next task is to establish two relations between m and w. Rice, soup and meat serving dishes give us one number: we know that 52 individual dishes were needed to serve the meal. The rice course used most dishes because only two people supplied themselves from each serving dish. The soup course used the next largest number, with three people serving themselves from each dish. The pork course used the least number, with four people serving themselves from each dish.

If we call the number of dishes R (for rice), S (for soup) and P (for pork), R is more than S, which is more than P. Three fractions now express the relation between the number of each kind of dish:

R 1 dish for 2 persons		S 1 dish for 3 persons		P 1 dish for 4 persons		Total number of dishes
$\frac{1}{2}(m+w)$	$+$	$\frac{1}{3}(m+w)$	$+$	$\frac{1}{4}(m+w)$	$=$	52.

Rice dishes are equal in number to half the group; soup dishes are equal to a third of the group; pork dishes are equal to a quarter of the group. Together, the number of dishes equals $\frac{1}{2}+\frac{1}{3}+\frac{1}{4}$ of the group, that is $\frac{13}{12}$ths. The number of dishes is one-twelfth greater than the number of persons. Therefore, since the total number of dishes is 52, there were 48 people at the feast. In modern terms:

$\frac{13}{12}$ times number of persons $= 52$
therefore $\frac{1}{12}$th times number of persons $= \frac{52}{13} = 4$
therefore number of persons $= 12 \times 4 = 48$

The first of our two simultaneous equations is therefore

$$m + w = 48 \tag{i}$$

At this point, with two unknowns (m and w) but only one equation, the solution is indeterminate. However, the cook reported that in the group game, there were equal numbers of men and women. In modern terms,

$$m = w \tag{ii}$$

Once we have two equations connecting the number of guests and the number of entertainers ($m+w=48$; $m=w$), we solve the problem by substituting the values from one equation in the other:

$m + w = 48$
therefore $2m = 48$
therefore $m = 24$
Then: $24 + w = 48$
therefore $w = 48 - 24$
therefore $w = 24$
There were 24 women (entertainers) and 24 men.

Replacing modern symbols with verbal descriptions (as the Chinese would have done), we reach the same result by the same reasoning, as follows:

2 people use 1 rice serving dish
3 people use 1 soup serving dish
4 people use 1 pork serving dish
Half of the group plus a third of the group plus a quarter of the group uses a total of 52 plates.
But if we add one-half to one-quarter and then add another third it equals more than one (group) – in fact, it is one-twelfth more: that is, it adds up to thirteen-twelfths. 52 can be divided into thirteen heaps of 4. 12 of such heaps would be 48.
So there must have been 48 people altogether.
The cook now informs us that there were equal numbers of men and women (shown by the group game), and we deduce that half the group must therefore have been men, and half women.
Therefore, there were 24 men and 24 women in the group.

This is the oral form known as 'rhetorical algebra'.

Indeterminate equations

The idea of the indeterminate equation would occur easily to a bureaucrat. This kind of equation was known as *tai yen* in Chinese. Its attraction lay in the fact that it models the traditional qualities of Chinese thought. We can solve the problem by abstract logic, but we have then to select from all possible solutions the one, or ones, that best fit the practical situation.

Chin's treatise describes one such situation. In a market, fowl are priced as follows: four cash for a cock, five for a hen and one for a chicken. The cook wants to buy some. He has 100 cash for the purpose, and wants to buy some of each kind. How many of each would you advise him to buy?

Chin's solution depends on the algorithm (obvious from the above information) that one cock and one chicken are equal in price to one hen. We can also see that if the cook bought one kind of bird only, his 100 cash would pay for 25 cocks or 20 hens or 100 chickens. The algorithm implies that each time we take away one hen we may increase the number of other birds: cocks by one and chickens by one. This keeps the total cost constant at 100 cash. The total number of fowl, however, alters according to the option chosen. Some of the options are shown in the table on page 61.

However, none of these suggestions satisfies the cook. He has 60 guests and must buy at least 60 fowl. This necessity limits his choice in other ways.

*Algorithm →	Cocks +1	Hens −1	Chickens +1	Total fowl (n)	Total cost
	10	10	10	30	100 cash
	11	9	11	31	100 cash
	12	8	12	32	100 cash
	13	7	13	33	100 cash
	14	6	14	34	100 cash
	15	5	15	35	100 cash
	16	4	16	36	100 cash
	17	3	17	37	100 cash
	18	2	18	38	100 cash
	19	1	19	39	100 cash

*The algorithm is: 'add one cock and one chicken for each hen taken away'.

One solution – not the only one, but the best if he is resolute on not spending more than 100 cash – is to buy 65 fowl altogether: 5 cocks, 5 hens and 55 chickens. $((5 \times 4) + (5 \times 5) + (55 \times 1) = 20 + 25 + 55 = 100.)$ This problem became a classic among Chinese mathematicians.

Modular arithmetic

The indeterminate equation is useful in solving other problems of everyday living. The original problem, in what we now call modular arithmetic, was classically stated by Sun Tzu in the 4th century:

We have a number of things, but we don't know exactly how many. If we count them by threes there are two left over. If we count them by fives there are three left over. If we count them by seven we have two left over. How many things are there?

A specific example will show not only how the problem might have arisen in real life, but also how it could be solved. Let us imagine that a magistrate has been asked by the tax department to look into the affairs of three farmers, A, B and C. They are brothers who work together, growing rice. When the crop is ready, they divide it into three equal portions. Each takes his share to a different market. They think that they get a better price that way. The problem is that the markets are in different provinces and use different units to measure the crop. When the farmers sell the rice, there is a small amount left over in each case. They bring this back and sell it in the village. A brings back two catties (a catty was the ancient Chinese unit of weight, approximately equal to 1 lb 5 oz), B three and C two.

The magistrate interviews the brothers. They claim that they actually sold 804 catties of rice but he refuses to believe them. He knows that three units in A's case were equal to five in B's and to seven in C's. He knows from their neighbours that A brought back two catties, B three and C two. He knows from the markets in various provinces that they had equal shares at the start. The problem is: how to prove that they are lying.

The key to the solution is the relative size of the measures. We begin by making them standard in terms of the different units. We do so by dividing by the ratios. We know (from the information supplied above) that $3A = 5B = 7C$. We therefore divide A's share by 3 (remainder: 2), B's share by 5 (remainder: 3), C's share by 7 (remainder: 2). The problem can now be stated in numerical terms: What is the single number, close to 268 (the alleged share, one-third of the declared crop) that is divisible by 3, remainder 2; divisible by 5, remainder 3, and divisible by 7, remainder 2?

The Chinese remainder theorem says that it is possible to find a number that will satisfy any given collection of divisors and remainders. (For example, a number that leaves the remainder 2 when divided by 5, 6 when divided by 7, and 3 when divided by 4, is 27.) However many conditions are stated, a number satisfying them can be found. To do precisely that is the magistrate's task on this occasion.

There is a simple algorithm to find such numbers. You begin with the given remainders, then add the relevant divisor until you have equal amounts. In the case of the three farmers this works as follows:

	A	B	C
Divisor.	3	5	7
Remainder	2	3	2
1st guess (remainders):	2	3	2
add divisors			
2nd guess:	5	8	9
add divisors			
3rd guess:	8	13	16
add divisors			
4th guess:	11	18	23
add divisors			
5th guess	14	23	

The same numbers have been reached for B and C; continue with A alone:

6th guess	17
7th guess	20
8th guess	23

So 23 catties is the smallest amount of rice that could have been grown by each farmer, to conform with the remainders sold in the village. But

$3 \times 23 = 69$ – far less than 804 (the amount claimed). The farmers are hardly likely to have overstated the amount on which tax is to be paid. However, if you add 105 (that is $3 \times 5 \times 7$, the product of the three divisors) to 23 (each farmer's share), the result (128) still satisfies the remainder conditions. The same is true however many times you add 105.

A	B	C	Total crop
23	23	23	69
128	128	128	384
233	233	233	699
338	338	338	1014

Using the indeterminate concept thus allows the magistrate to show that the farmers actually grew not the amount they claimed but 1014 catties. They therefore underdeclared their income, and are in trouble.

The calculations set out above use the 'blunt instrument' approach and modern notation. But the problem is just as easy – easier – to solve using a counting board. Sticks are set out in order at the start, and added or removed as each calculation is made, until only one line of sticks is left: the answer. The steps, with interpolations to show how the Chinese scholar's mind may have worked as he did the mathematics, can be set out as follows:

Step 1

The celestial units are	1	1	1
The original divisors are	3	5	7
Operation numbers are	35	21	15

The operation numbers are products of the divisors, taken two at a time. The remainders when 35 is divided by 3, 21 is divided by 5, 15 is divided by 7, are shown next.

Remainders	2	1	1

The celestial units are used to point the way to the operation numbers. These are essential, being the multipliers for the operation numbers. Our first guess is 2, 1, 1. The two 1's are good (in fact, they are the celestial units that we are looking for). Now we need only find the third.

Step 2

The procedure here was known as *t'ai yen chiu i shu* ('great extension method'). The working is shown below. We divide 3 (original divisor) by 2.

The result is 1, with remainder 1. This result times the celestial unit, 1, gives us the reduced number $1 (= 1 \times 1)$. We now have 1 (as celestial unit); 2 (as original remainder);? (as unknown number); 3 (as the base). Hence:

1 (celestial number) 2 (original remainder)
1 (reduced number) 3 (base)

Step 3
Now we divide the reduced number, 1, by the remainder of the base, 1. The result is 1. We multiply this by the reduced number and add the celestial number to make $2 (1 \times 1 + 1)$. This 2 is the number we have been looking for to serve as multiplier.

We now have 2 (as multiplier); 1 (as residue); 1 (as reduced number); 1 (as remainder of base).

Step 4

Operation numbers (from above)	35	21	15
Multipliers (celestial units)	2	1	1
Reduced use numbers (35×2 etc.)	70	21	15
Original remainders	2	3	2
Reduced numbers × remainders (residues)	140	63	30

Step 5

Sum of residues $= 140 + 63 + 30 = 233$
Operation modulus $= 3 \times 5 \times 7 = 105$
Subtract $(233 - 105) = 128$
Again: $(128 - 105) = 23$

Answer: This 23 is the least number that yields the remainders we need. It is the number we found by the other method. To solve the tax collector's problem we do the reverse: we keep on adding 105 to 23. Eventually this gives the answer 1014, as before.

The remainder theorem (known in China as long as 1900 years ago) refers to the situation where we have to discover an unknown number (for example, in the knowledge of some other person, as the true amount of harvest in the example above was known to each farmer but not to the tax official). For convenience, we shall call this unknown number P. We also know that there are three or more divisors (say x, y and z, or 3, 5 and 7 in the case of the three farmers). The divisors must be 'prime' to each other: that is, they must neither divide into one another nor have a common factor. When x, y and z are divided into P, they leave remainders of p, q and r (in the farmers' case these are 2, 3 and 2).

The remainder theorem states that there is a number P, which can be found precisely by combining the remainders and the divisors as in the calculations shown above. The problem can also be solved, much more elegantly, using the modern notation of modular arithmetic. This refers a number to a base or modulus.

In the farmers' case, using modular notation, we are told:

$$P \text{ (mod 3)} = 2 \quad P \text{ (mod 5)} = 3 \quad P \text{ (mod 7)} = 2$$

We can summarize all the Chinese solution above in one equation drawn from the Chinese Remainder Theorem. (Full stops in this notation are multiplication signs. Figures in brackets are multipliers.)

Remainder from $3 = 2$; from $5 = 3$; from $7 = 2$
Then $P = (2.(2).5.7 + 3.(1).3.7 + 2.(1).3.5)$
$\qquad = (140) + (63) + 30 = 233$

This gives one possible solution of the problem. If we subtract 105 twice, the remainder is 23 as before. In other words: 233 (mod 105) = 23. To find other solutions we just keep on adding 105 successively. This will always leave the three remainders as before, and the total will eventually reach the number greater than 804 that we are looking for.

To non-mathematicians, this may all seem hocus-pocus, the kind of number sleight of hand that irritates as many non-specialists as it enthrals. In fact, it is extremely simple – so much so that a form of it was used, precisely as hocus-pocus, in the 'Chinese mind-reading act' (performed by Chung Ling Fu?) in the old-time music hall. In this, the magician asks a member of the audience to think of any number less than 316 – not to reveal it, just remember it. Next, the magician asks the person to divide the number, in turn, by 5, 7 and 9, and to state the remainders. Once the magician knows the remainders, he or she can, with a flourish, reveal the number first thought of. This is done with showmanship, not brow-furrowed calculation. But calculation is still the secret. If the remainders are 4, 6 and 9, for example, the magician can quickly work out that the hidden number is 314. He or she could use exactly the same method as the Chinese magistrate, though employing memorised tables instead of calculation.

Chinese astrology: equations of high degree

The calendar was an essential instrument of policy in the Chinese empire. The smooth running of imperial affairs depended on the calendar, and the calendar depended on careful observation of planetary motion, and on accurate predictions of such things as eclipses. Precise timing, for example, was needed in the emperor's daily rituals which, his subjects thought, guaranteed order in nature and just government and prosperity throughout the empire. The Mandate of Heaven (continuing favour of the Supreme Power towards the Imperial house) remained in force only if all ceremonies were performed at the right times and in the right sequence.

Mathematics was also a main support of the dynastic interest. It ensured the best line of succession. Among the matters programmed by astrology was the proper order of the imperial harem. Besides the empress and three senior consorts, there were nine spouses and 27 concubines. In addition, 81 female slaves served as assistant concubines, and a large corps of (female) secretaries was at hand to record imperial orgasms. The emperor's favours were bestowed according to a rota. Every 15th night was reserved for the emperor to have intercourse with the empress. The next night was reserved for the three consorts in a group. The nine spouses, in a group, were assigned the next night. Then the 27 concubines were chosen in rotation, nine each night. Finally, over a period of nine nights, the 81 assistant concubines were serviced in groups of nine. On the 16th night, the sequence started again with the empress and proceeded until the end of the month of 30 days.

It is beyond belief that the emperor fulfilled this rota to the letter, engaging in intercourse with all the ladies of the harem in sequence in the manner suggested. We can only speculate about what went on, for example, on the nine successive nights spent with nine assistant concubines. But the object of the plan is clear: to procure the best possible succession. The rota ensured that the emperor lay with the ladies of highest rank on those nights nearest to the full moon. The empress was favoured by having two nights at the full moon, the consorts next, then the spouses and concubines and the assistant concubines last. The empress's Yin, or female force, would be most potent and able to match the emperor's Yang, or maleness, when the moon was full. The purpose of the other females was to nourish the potent male force of the emperor with their Yin so that the Yang was a maximum at the new moon.

The corps of secretaries also played a vital role. In Chinese astrology, the horoscope is timed from the moment of conception, not from that of birth. The harem secretaries attended the bedding of the imperial partners, noting with which partner intercourse took place and the exact time of orgasm. The disposition of planets at this precise time could be shown on a machine invented by Su Shu. This was called the Celestial Clockwork. Information would then be available for the imperial astrologer to cast horoscopes for any children born, to help select the prince successor. Normally, this prince would be a son of the emperor's chief wife, the empress. But in case the emperor might change his mind about his successor, the secretaries had to play safe and keep a complete record.

Making planetary observations, keeping and analysing records and casting the horoscopes all called for much mathematical ability, including that necessary to solve equations of high degree. For example, Chin quotes the following problem (which we also 'translate', by adding the signs and unknowns of Western notation):

IIII		T	IIII	II	IIIII	T	black sticks	(–) 4064256
			⊤⊤	T	III	II	red sticks	(+) 7632 (x^2)
						I	black stick	(–) (x^4)

This problem is an equation of the fourth degree:

$$-x^4 + 7632x^2 - 4\,064\,256 = 0$$

(Convention determined that the first term in an equation was written as a negative.)

Although a method for solving such equations became known in Europe only in the 16th century, the Chinese had been solving similar problems, as far as the tenth power of $x(x^{10})$, for centuries before this. Their solution was directly connected to the method of extracting roots. The absolute term (the plain number, in other words) was the key. (At this time, Chinese scholars did not know that equations above the first degree have, in fact, not one, but several roots. So they set out to find, as in the example below, one positive solution.)

The fourth root of 4,064,246 would seem to be a useful first attempt to solve the equation. An allowance has to be made for any middle terms (in this case, $7632x^2$). This involved the complicated procedure of reducing the equation by successively taking out figures as though calculating the fourth root of 4,064,256. (The method was rediscovered independently by Ruffino in 1804 and by Horner in 1819.)

The process started with a scrutiny of the absolute term. This led to an inspired guess at the first figure of the root. The number 4064 at the outset (the four figures beginning the constant term 4,064,256) suggests seven or eight as the first figure of the answer (7 to the power of $4 = 2401$; 8 to the power $4 = 4096$). Seven looks too small, so we try 8. Using the board, we put x equal to $(y + 80)$ and derive a new equation.

Original equation:
$$-x^4 + 7632x^2 - 4\,064\,256 = 0$$
New equation:
$$-y^4 - 302y^3 - 30\,768y^2 - 826\,880y + 3\,820\,554 = 0.$$

We now do a similar calculation for the second figure of the answer. We discover that ($y=4$) satisfies the equation above. When we substitute 4 for y, the equation vanishes. Therefore the solution that satisfies the original equation is:

$$x = 80 + y = 80 + 4 = 84.$$

This is the method proposed by Chin. Progress in algebra (Chinese and European) since his day has provided an alternative method which also finds the other factors. We can now deal with the original equation as if it were an equation in x^2, and solve it by using factors. The equation has 4 factors: $-(x+84)(x-84)(x+24)(x-24) = 0$.   It has therefore four solutions, not just the one proposed by Chin: $x = 84$; $x = -84$; $x = 24$, or $x = -24$.

Magic squares and numerology

The 'magic square' is an arrangement of numbers which can be added together in a great number of directions to yield the same total. According to Chinese legend, the Emperor Fu-hsi (29th century BC) was bathing one day when he found the footprints of a mysterious creature in the sand. They were identified as those of a 'heavenly dragon horse'. On a similar occasion, the Emperor Yu (21st century BC) encountered a 'divine tortoise' with mysterious markings on its back. In each case, the markings were taken at the time to be heavenly messages to the rulers about the principles of ordering the state. In later times (c. 5th century BC) the markings were interpreted as magic squares, of great magical, mathematical and spiritual interest.

For centuries after this identification, indeed into the present century, these squares were crucial elements of Chinese numerology, used in imperial ritual, by necromancers casting spells and as the basis for prophecies and horoscopes. Even the level-headed and practical Chin is not prepared to discount their importance:

There are conductors who can harmonise the sound of bells and musical stones. But we cannot say that they produce that harmony between Heaven and Earth which the Great Music is said to do. Some again harmonise Yin and Yang, the growth and decay of the seasons, the five notes of the scale, the pitch pipes used in divination.

Back of the divine tortoise	Footprints of the heavenly dragon horse
Numbers add to 15, vertically, horizontally and diagonally: $4+3+8, 9+5+1, 2+7+6;$ $4+9+2, 3+5+7, 8+1+6;$ $4+5+6; 8+5+2.$	Yin (even numbers) and Yang (odd numbers) add to 20: $2+4+6+8, 1+3+7+9;$ five is central and excluded from the count.

. . . But this Esoteric Mathematics (as it is called) cannot be separated from the minor art of earthly calculation.

(Shu Shu Chin Chiang, preface, freely translated)

It may seem a blot on Chinese mathematics that numerology was allowed such importance for so long – that, for example, the 7th-century Empress Wu, designing the Ming Tan (Hall of Light) in Peking, as a centre for the imperial rituals, should have based her calculations on magic squares, following superstitious rather than architectural necessity. In defence of the Chinese, it should be said that similar beliefs are common in all cultures. Number symbolism is a major part of superstition, even (or especially) in 'advanced' modern societies. Magic squares and other 'special' combinations of numbers have been used in divination, astrology and spells, and in connection with specific ritual objects such as the swastika and the Christian cross.

The Chinese value of pi

Pi (the ratio between the circumference and diameter of a circle) is an essential constant in many practical calculations, Its exact value is one of mathematics' Great White Whales. Pi is an irrational number: one that cannot be calculated precisely. The interest lies in discovering the most precise value, to the greatest number of decimal places. Over the centuries, several mathematicians have devoted their entire lives to the work, calculating the value to 20 places, to 30, to 100. In computer days, it has been worked out to a million places. Possibly as much attention has been paid to this pointless exercise as humans have spent in preparing to wage war on one another. It is a game mathematicians never tire of playing.

One plausible reason for concern with pi is that, like Everest, it must be conquered 'because it is there'. Calculating pi is an index of the level of mathematical knowledge prevailing at any given time and place. For example, the fact that the ancient Hebrews were satisfied with a value of 3 as their best estimate of pi (used in constructing Solomon's temple) indicates their lack of interest in precision with numbers. The ancient Egyptians had the value $\frac{22}{7}$, the same as that quoted by Pythagoras, and which we use today as the best approximation for everyday calculation. It served the Egyptians even in their large-scale land surveys and buildings works.

Chin quotes these imprecise values of pi, and was not averse to using them. He also gives the quite inaccurate value (also used by the Hindus) of the square root of 10, and the far more precise value, calculated in China some 1700 years before his time: $\frac{355}{113}$. It was not ignorance that guided his conduct, but convenience.

Chinese ideas and the computer age

The most direct link between ancient China and the modern computer age (forged by Leibniz and the Jesuit missionaries to China, especially Adam Schall SJ) was the idea of the binary system. This number system is one of the six or so most basic ideas behind the modern electronic computer. It is the basis of the codes underlying most computer operations, and the medium in which the computer performs its arithmetic function (that is, addition, subtraction, etc.: see page 232).

Another link, less direct but just as vital, was the Chinese idea of rendering counting processes mechanical by the use of a physical apparatus (counting board and sticks and later, the abacus). As a spin-off, the Chinese discovered many of the fundamental algorithms and concepts of algebra and arithmetic. These became obvious because of the relation between the mental analogues, or models, and physical reality. In the high noon of Chinese scholarship, calculation, despised in other cultures as being suited only to slaves and drudges, was carried out as an essential state duty by scholars who were personal advisers of the emperor. The rupture between theory and practice, an early and enduring feature of Western civilisation, was never characteristic of China. It must be considered as a mark of decadence, both social and intellectual.

The modern computer is a direct descendant of the Chinese infatuation with precise and complex work with numbers. No one can be in doubt as to the pre-eminent contribution made by Chinese scholars to this development towards precision. The decimal system, the place-value concept, the idea of zero and the symbol we use for it, the solution of indeterminate, quadratic and higher-order equations, modular arithmetic and the remainder theorem – all were taken up by European scholars millennia or centuries after they had been invented by the Chinese. The counting board became standard equipment in merchants' offices in mediaeval Europe as an indispensable adjunct to number work. Used in the sale of goods in shops, the 'counter' became the centre-piece, separating the customers from the salespeople. The abacus, too, became an essential piece of equipment – and indeed is still widely used in Eastern Europe and much of Asia.

1 2 3 4 5 6 7 8 9 10 11 12 13 14 15 16 17 18

ANCIENT GREEK FANTASIES ABOUT NUMBER

Ordinary Greek calculation remained to the last so clumsy and primitive,
that if any progress in the art is to be ascribed to the Greeks, it can be
exhibited only by going back to the beginning.   *James Gow*

We have outgrown the phase when all the arts were traced to Greece, and
Greece was thought to have sprung like Pallas, full grown from the brain of
Olympian Zeus; we have learnt how the flow of genius drew its sap from
Lydians and Hittites, from Phoenicia and Crete, from Babylon and Egypt.
                                                      Leonard Woolley

GREEK thought, in Latin dress, dominated Europe for sixteen centuries.
The reason for this was that the views of Socrates, Plato and Aristotle were
taken up, worked over and translated into Christian doctrine. In fact,
Christian thought is best described as a meld of Greek philosophy purified
of its overtly pagan elements, and Hebrew monotheism purified of its
Jewish exclusivism. Socrates, sentenced to death for corrupting the youth of
Athens, was identified by Christians as a prefiguration of Christ crucified.
Plato's account of his death (in the dialogue *Phaedo*) was regarded by early
Christian scholars as having the same status as the Bible account of John
the Baptist's ministry.

The Jewish heritage of Christianity was passed on by the first disciples,
Peter and the others. The Greek gift, which changed this Jewish sect into a
world religion, came from Paul. The mixture of Greek and Hebrew thought
was so potent that the Christian religion for many centuries filled the gap
left by the deaths of Greek paganism and Roman power. One outcome was
that the progress of mathematics in the Western world was seriously
impaired.

Objective understanding of our Greek heritage, in the context of other
ancient civilisations, is comparatively recent. It is only in the last 200 years
or so (beginning with the discovery of the Rosetta Stone, by a soldier in
Napoleon's army in Egypt in 1799, and its decipherment some 25 years

later) that we have been able to read ancient inscriptions other than in
Greek or Latin, and to take the intellectual measure of some of the cultures
whose relics archaeologists have unearthed in such abundance during the
same period.

One result of this work has been to show that, contrary to earlier notions,
Greece was not the fountainhead of the sciences. Those who gave this credit
to the Greeks were mistaken not only about the source, but also about the
nature of science. The Greeks made a unique, but relatively minor,
contribution to mathematical logic. They tidied up the body of knowledge
about geometry, making it into an abstract system knit together by
deductive logic. (This was an achievement of genius by Euclid: see page 85.)
They created a climate of opinion in which arithmetic was also considered
as an abstract system, with no practical application. But from the
standpoint of modern, practical science, these contributions had the same
kind of baleful, stultifying effect as the Greek views on society – for example,
on education or the status of women – which still bedevil our cultural and
social values.

Arithmetic and logistics

In modern times, arithmetic means something quite different from what it
meant to the Greeks. Scientists describe it as 'the language of science', a
major tool in the study of nature and society. The Greeks, however,
thought of it as a form of abstract wisdom, with no connection whatever
with practical activity.

Calculation is clearly a mental art. It is part of a battery of skills. Children
(and slaves) can be taught to use logistics in the form of drills (what we call
mechanical arithmetic), with a minimum of mental content. But it can also
be taught at the highest level, as in mathematical physics, and at all stages in
between. Skills are used by craftsmen and craftswomen to produce artefacts
of value and practical use. They relate to the real world. Physical skills
always involve some mental component, for example educated judgement
or the calculation of size and scale; even the most mental of skills (for
example, composing music) involve some physical activity.

In this day and age, these remarks are platitudes. But they would have
made ancient Greek 'thinkers' laugh out loud. Calculation (logistics), for
them, was a skill like fishing or barbering; arithmetic was a different and
higher form of art, whose practice was for free citizens only, forbidden by
law to slaves. This attitude applied even to numbers and numerals. To the
modern thinker, they seem identical. But to ancient Greeks they were
utterly different, their nature depending on their context and especially who

was involved in the operation. In arithmetic, numbers were regarded as abstract, spiritual entities; numerals in logistics (written exactly the same) were regarded as base, concrete quantities, with no 'existence' independent of the objects that they described. (Most 'moderns' would agree with this last statement.)

Greek scholars constantly made distinctions of this wrong-headed kind: education versus training, sport versus work, free person versus slave. These distinctions were definitions to guide the inquirer; they were linked to matters of life-style. They were also blinkers, hiding the real connections between things, and distorting every subject to which they were applied. In the 6th century BC, for example, Parmenides argued that whatever exists cannot have come into being. If it had, it must have come either from itself or from not-itself. But both of these are impossible, because according to the law of non-contradiction a thing cannot be itself and not-itself at the same time. In such a discussion, there would have been no point in drawing attention to such realities as sexual intercourse (which may initiate a process of becoming). It was a cardinal principle of this logic that experience was ruled out (since it came to us not from reasoning but through our senses). Experience was not allowed to correct, or even to throw light on, the process of reasoning. Indeed, it was the main function of reason to correct the false conclusions drawn from the errors imposed by our partial view of reality.

Using the method of reasoning illustrated above (a method which he invented, known as the *reductio ad absurdum* or method of indirect proof), Parmenides and his disciple Zeno demonstrated not only that change was impossible, but that motion was a logical contradiction. There was neither becoming nor perishing, neither space nor time. If Achilles ran a foot-race against a tortoise, where the latter was given a start, he could never overtake it, because each time he came to where the tortoise had been when he started out, it would have moved a little way further on. An arrow shot into the air must remain suspended in the same place, forever (see page 180). Such conclusions may be contrary to experience and the common understanding of humankind, but they can be proved by reason. Since, by definition, reason is the only guide to truth, we must reject the evidence of our senses. Practice is not a guide, nor is it relevant as a corrective of theory. In fact, the best reasoning is that which can be shown to contradict practice, as in the above examples. It reminds me of the attitude of the naval commander who ordered 'Damn the torpedoes! Straight ahead!' or the World War I sergeant encouraging his troops: 'Advance and attack the guns. Do you want to live for ever?'

In such a system of reasoning, distinctions are made absolute; they achieve the status of definitions. Thus, reason told the Greeks that training was totally different from education. It was provided for slaves and for children: suitable clients because their mental powers were not fully developed. It was physical rather than mental, directed towards production rather than to bring those instructed in touch with ultimate reality (the purpose of studying number theory – another name for 'arithmetic').

Greek thinkers and the natural world

Although scientific study, in any modern sense, was outlawed by such views – it dealt with practicalities not concepts; not with absolutes but with the observation of change – there was a rival scientific tradition in Greek thought, concerned with understanding the natural universe and the place of human beings in 'reality', without reference to the gods. Certain Greek thinkers who preceded Plato, for example Democritus and Heraclitus, spent much time arguing about which 'elements' made up the universe. In the course of this discussion, they assumed that the world and its history were part of a normal, not a supernatural, process. They believed that something external to mortals and gods (what we now call 'matter') was the original substance from which everything developed. Events were the result of a cause-and-effect relationship which underlay all phenomena. The first loyalties of the human race were to truth and to itself. These were answers to the most basic questions of philosophy.

Various ideas were proposed about the first form of matter. Thales was the first thinker ever to attempt to explain all the variety of nature by something within nature and not outside it. He said that it must be water, the only substance he knew that could be observed to change (from the solid state, ice, to the liquid, water, to the gaseous, steam).

Other thinkers proposed different 'prime elements'. Anaximenes said air; Heraclitus said fire; Xenophanes said earth. Some argued for all four elements together. Some imagined a fifth element, or 'quintessence': aether, a subtle emanation and blend of the other four. Following the rule that concrete and abstract must be separated, all thinkers believed that the prime elements, whichever they favoured or in whichever combination, were different from their everyday forms, not the same as ordinary earth, air, fire and water.

This whole discussion was really pseudo-scientific, leading nowhere. Questions about the number and nature of 'first elements' were premature in the state of knowledge available at the time. But other problems in natural history were studied by observation and, in some cases, by experiment. Democritus (4th century BC, a contemporary of Plato) put forward an atomic theory and gave a modern explanation of how the senses worked in human beings. Aristotle, three generations later, did practical research, including dissection (abhorrent at the time) for his writings in biology. Archimedes (3rd century BC) was more like a modern scientist than anyone else then living.

None the less, all such scientific speculation and inquiry was subjected to unremitting attack by Plato and his followers (who put many of their arguments into the mouth of the dead but revered Socrates, so giving them a kind of validation by association). The basic charge they made was that the study of nature (and of human beings as part of the natural order) could only result in opinions, not knowledge. Being based on sense-perception and experience (instead of reason, that is, definition of terms and logical analysis), such studies must be vain as far as truth (that is, certain knowledge) was concerned. For example, Plato dismissed astronomy because it studied the planets instead of the eternal verities of truth and beauty. He advised astronomers to give up direct study of the Sun, Moon and planets in favour of thinking about them as the products of divine creation.

In making such suggestions, Plato and his followers were taking it on themselves to tell scientists to abandon the method of science (that is, the observation of specifics) in favour of broad assertions based on deduction. They had no understanding that knowledge is furthered not only by deduction (logical reasoning from an initial premise) but also by induction (checking assumptions and reasonings against observed phenomena). But this lack of understanding did not prevent their view of 'knowledge', of the quest for understanding, taking a firm hold both on Greek thought, and on those who followed it, for many centuries. One fall-out for the study of Greek science and mathematics is that very little survives of the writings of those whom Plato regarded as hostile. Although we have what seems to be an almost complete set of the written throughts of the Socrates–Plato–Aristotle school, less than one-tenth survives of the scientific legacy of the Pre-Socratics, and practically nothing of Democritus. Heraclitus survives in a Roman work written in verse and covering a large field of scientific knowledge. Democritus, however, comes to us only in the form of quotations by his enemies – quotations usually selected by them, without context, for a polemical purpose. The idea of submitting his works to ordeal by bonfire is proposed and discussed in one of Plato's dialogues. It seems hardly unfair to conjecture that, at some point, the Platonists had just such a bonfire of the works of their opponents.

Greek numbers

The Greeks had two different number systems. The earliest, used on inscriptions recording public accounts in the 5th to 1st centuries BC, are known either as Herodian numbers (after the 2nd-century writer who described them), or as Attic numbers (after Attica, the area round Athens where they were chiefly used). They were on a base of 10. They used the initial letters of the number words to distinguish the different ranks or levels, and the number of signs used was limited. There was no zero. Numbers were written as in the Roman system, repeating the character if necessary, for example:

Ι	Δ	Η	Χ	Μ	ᒌ	ᒌ	ᒌ	ᒌ
1	10	100	1000	10,000	15	50	500	5000

ΧΗΗΔΔΔΙΙΙΙ	ᒌΜΧΧΗΗΗΔΔΔΔ	ΗΗΗΔΔΙ
1 2 3 4	62,340	3 2 1

The second system of numerals, known as the Ionic or Alexandrian system after its main areas of use, was created for calculation, and had entirely replaced Attic numbers by the 1st century BC. It consisted of the 24 letters of the Greek alphabet, and 3 archaic letters, each of which had an assigned value.

α	β	γ	δ	ε	ς	ʒ	η	θ
1	2	3	4	5	6	7	8	9
ι	κ	λ	μ	ν	ξ	ο	π	ϙ
10	20	30	40	50	60	70	80	90
ρ	σ	τ	υ	φ	χ	ψ	ω	λ
100	200	300	400	500	600	700	800	900

By adapting the notation in appropriate ways, fractions and ordinal numbers could be worked with. But although the system was simple to understand, it was still too complex for everyday use. There were too many signs, and there was a plethora of relations between the 27 letters. It is probable that calculations were made with beads as counters, either on a sand table, or strung on wires in an abacus (*abax* is Greek for 'table'). Greek alphabetic numbers were used in Europe until supplanted by Roman figures in the 10th century.

Pythagoras

Pythagoras was born on the island of Samos in 580 BC and died at Metapontum in Italy in 500 BC. As a young man he settled in Babylon, where for 20 years he studied and taught astronomy, mathematics and astrology. In 525 BC he went to Croton in southern Italy, where he set up a secret society (religious, not mathematical) devoted to exploring the mysteries of number. The community was ultra-conservative and authoritarian, vegetarian and ascetic. It was not a 'brotherhood' (as it has often been called), but a community of families. It may be the model for the ideal society described at length by Plato in the *Republic*.

The Pythagoreans believed in five basic ideas – ideas that have had a powerful influence on the thoughts of humanity ever since. These were: (i) The universe was created, and continues to exist, on the basis of a divine plan. The ultimate reality is not material but spiritual; it consists of the ideas of number and form. The ideas are divine concepts, superior to matter and independent of it. (ii) God created souls as spiritual entities. The soul is a self-moving number which passes from body to body (animal as well as human). Souls are eternal. They inhabit bodies for a limited time only, then pass on to a new existence in another body. Purification, based on a strictly moral and intellectual life, eventually releases the soul from the endless cycle of the 'wheel of life' to a perfect union with the Divine. (iii) There is an inner harmony and order in the universe. This results from the union of opposites. There are ten fundamental opposites which interact with each other, each pair performing a creative function. These opposites, the foundation of the world as we know it, are: odd/even; male/female; good/evil; wet/dry; right/left; rest/motion; hot/cold; light/dark; straight/curved; limited/unlimited. (iv) In human relations, friendship and modesty are the most important principles. Men and women should live an ascetic life in a sharing group devoted to the rearing of children, in harmony with the divine plan. This entails active devotion to the study of number. (v) The divine ideas, which created and maintain the universe, are those of number. The study of arithmetic is therefore the way to perfection. By devotion to study and to the rules of the sect, the individual discovers ever-new aspects of God's plan and the mathematical rules by which the universe is governed. This last principle was the most important of all.

Members of the commune were expected to exercise restraint and decorum in carrying out their duties. Celibacy was practised by the leaders. Women seem to have been accepted as equals in all activities of the group, a rare thing in the ancient Mediterranean world. Among other things, they taught mathematics, which suggests that they were educated to a standard also uncommon at the time. All property was held in common.

Later Greeks believed that Pythagoras invented arithmetic. He was certainly the person who, above all, gave to Greek arithmetic its characteristic emphasis. His theory of number, and his distinction of logistics from arithmetic, held sway for centuries. But there is little clear evidence about his own discoveries. Doctrines of the sect, and the sacred knowledge, were passed on by word of mouth under an oath of secrecy, as religious mysteries. Written accounts were forbidden; members were under oath not to speak of the secret lore. Indeed, it is said that Hippasus, a member of the group, was drowned for giving away the secret that some numbers were 'irrational'. (That they could not be written down exactly was supposed to be especially sensitive information. But the story is almost certainly false, as is the rationale that the 'secret' undermined the whole doctrine of the sect: that the universe was fully rational, being made of numbers which are in the mind of God. In fact, Pythagoreans dealt with the problem by explaining that only integers, that is whole numbers, were numbers.)

Pythagoras is credited with two mathematical discoveries above all others: the theorem in geometry which bears his name, and the description of the effect of the length of strings on the harmony or dissonance of musical notes sounded altogether.

There is no possibility that he originated the famous theorem, that 'in a right-angled triangle, the square on the hypotenuse is equal to the sum of the squares on the other two sides'. It was known in China, from land surveying, and in Egypt, from pyramid-building, centuries before he was born.

So far as the music is concerned, there is no evidence to show whether Pythagoras discovered the relationship between lengths of string and musical sounds, or was merely given the credit for it. Certainly, it was claimed for him – and it has been described as one of the two really important achievements of Greek experimental science (the other being Empedocles' demonstration that air is a substance that occupies space and is compressible). Pythagoras' discovery is of interest to us for another reason: unlike many of 'his' ideas about number, it refers to external nature and can readily be verified.

Notes in harmony with one another are produced by plucking strings that are related to each other in length in a simple way. The most harmonious notes are produced when the length of one string, set in motion, is a simple fraction of the other. The shorter the string, the higher the pitch of the note. (Middle C vibrates at a rate of 256 vibrations per second.) If we halve the length of a string by fingering, a note one octave higher is heard. If we double the length of the string, it sounds the octave below. Octaves have the simplest possible relation to each other, their lengths being in the ratio 1 : 2; they combine to give the most harmonious sound possible – or almost: two single strings of the same length, sounded together, merge their sounds, which are in the ratio 1 : 1. With lengths related in the ratio 2 : 3, the next most euphonious interval, the fifth, is sounded. If the ratio is 3 : 4, the next most pleasant, the fourth, is heard. However, with lengths in the ratio of (say) 163 : 173, the sound will be an unpleasant discord.

Pythagoras concluded from observations such as these that some notes 'go' together because the numbers that represent their lengths are related to each other in a simple way. Some notes form 'bad' combinations because there is no simple relation between the numbers that represent them. Similarly, he and his followers believed that the qualities of all things, whether concrete (for example, musical sound) or abstract (for example, justice), could be explained by number. Each was thought to have its own soul or essence, and passed it on to other numbers as it combined with them. It was like the mating of the gods.

Some numbers were friendly and compatible, and came together more easily than others. Some were wholly 'evil', did not belong with other numbers, and brought bad fortune to humankind. Odd numbers were female, even ones male. Males could unite with females, but there was no barrier preventing males from coupling with males or females with females. It was the task of arithmetic to discover all the different kinds of numbers, how they were related to each other, and their place in the divine plan. There was, in short, a theology of number, and the mathematician was a theologian who discovered and proclaimed the divine order.

The taxonomy of number

Such Pythagorean views raise the whole question of definition. What do we really mean by 'number'? The Pythagoreans gave three definitions: number is 'limited multitude'; it is 'a combination or heaping up of units'; it is 'a flow of quantity'. Because numbers were divine archetypes, in God's mind from the beginning, the study of arithmetic was, literally, a way to 'figure out' the divine plan.

Accordingly, reality is not made up of things; they are but pale reflections of the divine thoughts. In fact, only ideas are real. We live, Plato said, in a cave as it were, where outside events are shown as a shadow play, on the walls. We cannot see out of the cave and can never know the real world as God knows it. We – or at least the intellectual and moral (male) élite among us – must piece together the nature of reality, reasoning from the moving shadows on the walls of the cave.

To the Greeks, the numbers one and two were so special that they were more than numbers, not really numbers at all. Aristotle explains this by pointing out that a number is a heap or multitude of units. The unit is therefore the measure of the number. For example, 'five' means five units. But the measure of a thing cannot be the same as the thing itself: one, the measure, cannot be the same as the thing it measures. Therefore it is not a number. It is the beginning of a number series. Similarly, two is the start of the even numbers; by the same logic, it cannot be a number either.

For Pythagoreans, the monad (one) was the first number to be created. It was identified with God the creator in a special way (like Adam in the Old Testament, it was made in God's image). It had some divine qualities, such as unity and wholeness, creative potential and priority. In other words, God was the First Mover, Number was 'first' in the natural order, according to Plato and Pythagoras. To some extent the number two shared these special qualities, because it came first in the even number series.

The natural sequence continues with three and four. If we add these four numbers together $(1+2+3+4)$ we get 10 as the result. This marks a break. The sequence starts all over again on a higher level, that is 11 (ten plus one), 12 (ten plus two), etc. The sum $(1+2+3+4)$ had a special name in Greek – *tetraktos*. Its result, the Divine number, 10, was the innermost secret of the Pythagorean sect. On admission, each member had to swear on this number not to reveal the secret lore of the group.

The first division of number is into odd and even. An even number is any number that can be divided into equal parts; an odd number cannot be so divided. Those that can be divided evenly only once by two are called even-odd numbers. Even-even numbers are those that can be divided again and again by two until unity (one) is reached. Odd-even numbers can be divided by two and again by two, but we must stop before unity is reached. Odd-odd numbers are the product of two odd numbers. Examples of the odd-even classification of number are given below.

Even	4 6 8 10 12	Odd	3 5 7 9 11
Even-even	4 8 16 32 64	Even-odd	6 10 14 18 22
Odd-even	12 20 28 36 42	Odd-odd	9 15 21 27 33

The main result of this classification was to draw attention to the fact that some numbers could not be reduced: they could be divided only by themselves, not by 2, 3 or anything else. They were prime numbers. The others, being composite, could be divided up into their components, called factors. Some examples are:

Prime numbers 3, 5, 7, 11, 13, 17, 19, 23

Factors: $6 = 2 \times 3$
$\quad\quad\quad 44 = 2 \times 2 \times 11$
$\quad\quad\quad 63 = 3 \times 3 \times 7$

Composite numbers lend themselves to all kinds of manipulations. By studying the factors of which they are made up, we can find all sorts of relations between them. For Pythagoras, these relations testified to the fact that the Creator had a definite plan, which could be discovered by correct understanding of the number sequences. For example, the acolyte might break a given number into its factors and then add these factors together (omitting the number itself, but including unity as a factor). As an illustration, 6 equals 2×3; and 1 and 6 are also factors. We ignore the factor 6 (because it is the number itself). We note that 6 also equals $1 + 2 + 3$. So it is a 'perfect' number (which is a moral judgement, with no place in number theory). There are only a few cases where the factors add up to the original number. Pythagoreans called these 'perfect' numbers, and could find only four between one and 10,000:

$$6 = 1 + 2 + 3$$
$$28 = 1 + 2 + 4 + 7 + 14$$
$$496 = 1 + 2 + 4 + 8 + 16 + 31 + 62 + 124 + 248$$
$$8128 = 1 + 2 + 4 + 8 + 16 + 32 + 64 + 127 + 254$$
$$+ 508 + 1016 + 2032 + 4064$$

Nicomachus, a disciple of Pythagoras, wrote that, as with numbers, so with human qualities. Perfection is rare in numbers; goodness and beauty are rare in humans. Imperfect numbers are common; so are evil and ugliness. Mathematically speaking, imperfect numbers show various kinds of irregular and unbalanced structures, which can be discovered from their factors. By definition, imperfect numbers are those where the sum of their factors is greater or less than the number itself. Like monstrous births, such numbers have too many or too few limbs or organs. If the original number is less than the sum of its factors, it is said to be abundant (as, say, 12 whose factors 1, 2, 3, 4 and 6 add up to 16). If greater, the number is said to be deficient (for example, 8 whose factors 1, 2, and 4 add up only to 7).

Like much of Greek number theory, such distinctions lead only to sermonising. Nothing else follows. The next division, however, into types, is more promising. This marks the difference between 'amicable' (that is, 'friendly') and 'unfriendly' numbers. Here we compare the sum of the factors of two different numbers. If the sums are equal, the numbers are said to be 'amicable'. They have the same parentage and, at least in their own ideal world, could be expected to be more congenial than numbers that do not 'add up'. Pythagoras gave one such pair:

$$220: 1+2+4+5+10+11+20+22+44+55+110=284$$
$$284: 1+2+4+71+142=220$$

Other pairs include 17,296 and 18,416; 1184 and 1210. More than 1000 such pairs of amicable numbers are known. We can also recognise chains of 'sociable' numbers. Here three or more numbers are related, the sums of their factors being the same. If there were more than three, we might call them crowds, but none have so far been identified.

In geometry, Pythagoras is remembered for more than 'his' theorem, dealing with the mathematical relationship between the squares which can be drawn on the sides of a right-angled triangle. He wedded arithmetic to geometry in other ways, showing, for example, that numbers can be shown as geometrical figures. If we lay out pebbles, one for each unit, we can see that one pebble can stand for a point, two for a line, three for a triangle and four for a square. The progression can be continued; for each number there is a geometrical figure: pyramid (five), cube (six), icosahedron (seven), dodecahedron (eight), and so on. Parts of this table take Pythagoras, and Greek geometry, briefly into three dimensions. They are ways of representing the 'perfect solids': those in which every face is the same regular shape, and which can be enclosed in a sphere touching every corner. (The numerological implications of this, for Pythagoreans, are obvious.) The solids are pyramid (tetrahedron: four equilateral triangles), octahedron (six equilaterial triangles), cube (six squares), icosahedron (twenty equilateral triangles) and dodecahedron (twelve regular pentagons).

The discovery was important from the point of view of number theory, as many relationships that remain hidden from pure thought became obvious in the visual mode. Pythagoras went further. According to Plutarch, he identified each of the four elements (earth, water, fire, air) with a solid figure. (Such figures are well known to the mineralogist in the form of crystals – rock salt, for example, is grouped in the cubic system as it assumes the form of a cube when salt-water evaporates; so too for other shapes.) Pythagoras thought that earth had the form of the cube, water of the icosahedron, fire of the pyramid and air of the octahedron. The universe itself had the form of the dodecahedron. These identifications bear no relation to reality.

After Pythagoras's death, the commune continued to function for a time. His wife Theano and his daughters helped to maintain the traditions he had established. It is said that Theano did original work on the 'golden section' – a topic that became important into the theory of art many centuries later – but there is no record of this research. The school soon divided into two groups, those who were caught up in mysticism and rituals (*akousmatikoi*, 'those who hear') and those interested in advancing their knowledge of number (*mathematikoi*, 'those interested in science'). In a violent democratic revolution in southern Italy, many members of the community were killed and the group split up. The *mathematikoi* moved to Tarentum, the *akousmatikoi* left to become travelling mystics.

Pythagoras' views continued to influence not only Greek philosophy but the whole universe of thought in the West until a late period. Plato's theories of the soul, his account of the creation and properties of things as arising from number, were direct borrowings from the Pythagoreans. Pythagorean numerology influenced Christian mysticism. Galileo was considered by his followers to be a Pythagorean; Copernicus and Leibniz claimed to be in the same tradition. Newton was stimulated by Pythagoras' ideas and devoted a large part of his time to alchemy and similar pursuits under the influence of Jakob Böhme, a Pythagorean mystic.

Euclid

Euclid was born about 330 BC, probably in Alexandria where he was later involved in teaching. He wrote *Elements*, a book on geometry which displaced all other texts on the subject for more than 2000 years. It consists of 13 sections, which exhaustively dissect its subject according to a systematic plan that includes setting out the results of each step in a formal statement. His other surviving works include *Phenomena*, on astronomy, and *Optics*. He died in about 260 BC.

Elements begins with the definitions of a point, a line, a surface, a circle and parallel lines. Euclid then sets out five common notions ('axioms') which cannot be proved but which can be accepted as true by intuition.

They provide the necessary foundation for reasoning in this area. There are also five 'stipulations', which are similar to axioms, being also assumed to be self-evident, and provable not by logic but only by action. The main body of the work consists of a series of 'propositions': true statements built up by deduction in a systematic logical process from the axioms and stipulations.

Euclid makes no attempt to illustrate the truth of propositions by referring to cases from external reality, or to any practical application – an odd omission, considering the origins of his subject (the word 'geometry' literally means 'earth-measuring', that is, 'surveying'). His method of proof is entirely deductive. He regards geometry as a closed system of logical argument, requiring no input from human experience. It is concerned only with achieving the certainty that accompanies the discovery of truth. Apart from the fact that Euclid makes no reference to the Deity, his *Elements* might have been written by Plato himself. It imposes what are often rather senseless and arbitrary rules, such as the prohibition against using any instruments other than compass and straight edge to draw diagrams.

Unfortunately, the critical spirit was absent from geometry for many centuries, from the publication of *Elements* until about AD 1800. For good or ill, Euclid's treatise influenced the whole of mathematics in a way that was unparalleled by any other book during that time. With his 467 propositions, logically arranged in terms of the dependency of each later theorem on the earlier ones, with his insistence on the maximum economy of means in developing the logical argument, Euclid established his pre-eminence in this area. *Elements* was regarded as a perfect work, almost with divine authority. It was a brave spirit who would question this judgement, pointing out, for example, that the axioms were flawed in many ways, or that dozens of undeclared assumptions were made on the basis of Euclid's own unanalysed intuitions. *Elements* was also a compilation of earlier works, marred by repetitions, redundancies and illogicalities. For example, three sections on the theory of numbers (Books VII, VIII and IX) appear as if from nowhere, repeating for whole numbers (integers) results that have been already obtained for geometrical figures. These books also deal with prime numbers, geometrical progressions and perfect numbers.

In the 1820s and 1830s, Jonas Bolyai, a Hungarian army officer, Nikolai Ivanovich Lobachevsky, a Russian professor, and Carl Friedrich Gauss, the leading German mathematician, independently proposed a new kind of geometry, called non-Euclidean geometry. They each started from axioms different from those of Euclid (changing, for example, the axiom 'that parallel lines never meet'), and produced a variety of different results. Their work has only recently begun to be assimilated into the body of 'accepted' mathematics; it is still taught only in a few prestigious universities, and remains as a kind of professional secret. The fact is that Euclid's geometry was limited to a flat, two-dimensional plane (apart from Books XI, XII and XIII, which describe solid geometry dealing with three dimensions at right angles to each other). It is a 'flat-earth' view of reality. Non-Euclidean geometry provides a circular 'earth-based' geometry where we work with curved planes (as on the outer or inner surfaces of a sphere).

Euclid's overall philosophy and method of working are hostile to innovation in general, and to 'new' and computer mathematics in particular. It is not, perhaps, surprising that his thinking should have been so limited – why should he, more than anyone else, be free from error and bias? What is extraordinary is that his ideas were regarded as sacrosanct by so many and for so long, when evidence of the senses should have shown people that mathematics, and the world, were not like that at all.

Greek algebra

Algebra is an extension of the rules of arithmetic to discover the values of unknown numbers. This is done by using a known function to relate the unknown to some definite number. In the history of algebra there is a progression from 'rhetorical algebra', which uses mainly words and which was known to ancient Chinese and Egyptian mathematicians, to 'syncopated algebra', which also uses words, but in which special symbols enable us to economise on the rhetoric.

Numerous examples of syncopated algebra appear in the *Greek Anthology*, a collection of songs, poems, epigrams and riddles from the 7th to 4th centuries BC. The riddles were probably ancient in origin, and are in an area now called recreational mathematics. An example will show the level of argument. A number of apples is to be divided between six people. The first person gets $\frac{1}{3}$ of the total, the second $\frac{1}{8}$, the third $\frac{1}{4}$, the next $\frac{1}{5}$. The fifth gets 10 apples, and only one apple is left for the sixth person. How many apples are there altogether? (Answer: 120.)

Other writers on number theory, such as Iamblichus, Theon, Heron and Archimedes, sometimes engaged themselves with problems in nature that were algebraic. But there was little or no system in their work; they failed, for example, to develop a body of knowledge to compare in depth or rigour with Greek geometry.

With the appearance of Diophantus' book *Arithmetika* (3rd century AD), however, the situation changed. Nothing is known for certain about Diophantus except that he probably lived in Alexandria. Apart from *Arithmetika*, his books include one on porisms (additional conclusions that

can be drawn from theorems in geometry, now usually referred to as
'corollaries'), one on fractions and one on polygonal numbers (numbers
represented by dots which build up geometric figures: see page 84 above).
Even in the case of *Arithmetika*, his major work, only six out of thirteen
sections have survived.

In *Arithmetika*, Diophantus treats in detail various kinds of what we
would now think of as algebraic problems, and solves them in an algebraic
manner. He was probably aware that his work was new, but made no
general statements about the subject. He seems to have regarded it as a
special branch of number theory, designed to discover unknown numbers
that satisfied specific conditions. He builds on the algebraic identities in the
seventh and tenth books of Euclid's *Elements*, but (except for one section
which deals with the sides of a right-angled triangle) makes no use at all of
geometry to solve or illustrate the problems posed.

Diophantus treats 189 problems, not dwelling on abstract principles or
illustrating general rules, but merely giving a specific solution for each
problem posed. It may be that, like Euclid's *Elements*, his book is a
compilation of problems inherited from more ancient sources. His solutions are extremely ingenious, in a unique style which seems to mark them
as Diophantus' own.

The following problem, and its solution, are typical of Diophantus'
work. The problem is to divide a square number into two squares
(fractional solutions are allowed). In modern form, Diophantus' solution is
as follows:

Let the given number be 16
Let x^2 be one of the required squares
Therefore the other must be $(16 - x^2)$
Now take a square of the form $(mx - 4)^2$
Let $m = 2$
Then write $(2x - 4)^2 = 16 - x^2$
Expand left-hand side: $(2x - 4)^2 = 4x^2 - 16x + 16 = 16 - x^2$
Therefore we must have $5x^2 = 16x$
Divide both sides by x
This gives $x = \frac{16}{5}$
Squaring this gives one number: $\frac{256}{25}$
The other must be $\frac{144}{25}$
Both are squares, and add together to give 16

We can now put $m = 3$, or 4, or 5 or 6 . . . or each of these in sequence. Each
time we do so, we come up with a different solution: $\frac{1024}{289}$, $\frac{3600}{289}$ and so on. In
other words, there is a large number of possible solutions to the problem. It
belongs to the category we have described as indeterminate equations. Such

equations, and the method of solution, are characteristic of Diophantus. In fact, until the work of ancient Chinese mathematicians in this area was discovered, he was credited with inventing them.

Certain features of the above solution should be mentioned. First of all, it is most un-Greek in that it deals with a particular case (that is, it limits the method by taking the given number as 16, for example). In developing a subject, a Greek thinker would normally state a general case and deduce all possible solutions. The new variable m is introduced, so that we have a general formula which we can use (without mentioning actual numbers at all, but working out a general formula or routine) and then deduce all possible solutions. It plays a role similar to the rule of false position in that it de-centres the problem. It allows us to insert different values (for m) and thus to confer a degree of generality to the answer. But Diophantus does not take this as far as substituting a general symbol, like 'n', in place of the given number, 16.

In solving Diophantine equations we use the ordinary processes of arithmetic: we calculate roots and powers, and use the rules for the expansion of the squares of sums and differences. The only solutions we accept are whole numbers or fractions. Diophantus was also the first person to introduce symbolism into Greek mathematics (following the Babylonians): he used the Greek terminal letter sigma ('ς') to refer to the word *arithmos*, 'number'. It was characteristic of his method, if there were two unknown numbers, always to start by expressing one in terms of the other. The only other symbol he used, the sign for subtraction, had previously been used by Hero of Alexandria, two centuries earlier.

Because of the practical nature of the problems Diophantus set (to do with sums of money, for example), the Greek mathematical establishment never gave him proper credit for his originality. His contribution was regarded as no more than a degenerate form of number theory. He still stands in the vestibule, or antechamber, of symbolic algebra. He is the captive, and to some extent a victim, of the Pythagorean–Platonic attack on empirical science.

Symbolic algebra: the Greek contribution

As one certain of all his opinions, believing that once merely stated, they must prevail, Samuel Johnson once remarked: 'Greek, sir, is like lace; every man gets as much of it as he can'. Clearly, Johnson did not learn his mathematics in the classical mode. If he had, he might have had a very different opinion of Greek letters, at least when used as numbers. There is hardly a need to argue this case. It should be enough to set out a

Diophantine problem in modern notation, and in Greek notation, and let the differences between them speak for themselves.

The problem is to find two numbers such that their sum, and the sum of their squares, are two given numbers. Picking our steps not quite at random, we say: let the given numbers, when found, sum to 20; let the sum of their squares be 208. In the style of Diophantus, we then express the two required numbers in terms of one (not two) unknowns, 10 plus x and 10 minus x. From Euclid Book VII, we know that, if we square the sum of these two numbers, we obtain, to begin with, two squares with sides equal to 10 and x respectively. This is true in both cases. In addition, in the first case, we have to add on two rectangles each with sides of length 10 and x; in the second case, we have to subtract these two rectangles from the sum of squares. If we now add these results, the two rectangles added in the first case are cancelled out by the two we take away in the second case. When we subtract 200 from both sides of the equation we are left with twice the square of one of the desired numbers. This makes the other number easy to find.

This problem and its solution are set out symbolically, in modern notation and in Greek style, below.

English translation

Let the sum of the numbers be 20
  and the sum of the squares be 208
Let the numbers be $10 + x$ and $10 - x$
Squaring gives $x^2 + 20x + 100$
              $x^2 - 20x + 100$
Adding for sum of squares gives $2x^2 + 200 = 208$
Subtracting 200 from both sides gives $2x^2 = 8$
This must mean that $x^2 = 4$
And so the solution is that $x = 2$

Hypatia

When Alexander the Great invaded Egypt in 332 BC, he founded the city of Alexandria at the mouth of the Nile. Within a century it had a million inhabitants and had replaced Athens as the cultural centre of the Hellenes. Among the professors who taught at the famous Museum (university) there was the mathematician Theon. He became well known because of his work on Euclid's *Elements* and Diophantus' *Arithmetika*. He was a Pythagorean, and made himself responsible for the education of his daughter Hypatia (at a time, 700 years after Pythagoras, when it was comparatively rare for women to receive any sort of intellectual education at all).

Hypatia cooperated so well in this enterprise that she was appointed professor of mathematics and philosophy in the same prestigious institute of advanced studies as her father. Following in his footsteps, she was a pagan, a Platonist and a Pythagorean. Her courses on mathematics were particularly popular, attracting many foreign students. As is usual with women professors in the ancient world, reports of her beauty were supported by tales of how she lectured from behind a screen which concealed her beauty from her auditors and saved them from distraction.

By Hypatia's time, the Museum, and knowledge generally, had been in decline for many years. Since Ptolemy's death two centuries before, there had been few significant advances in mathematics or in the natural and physical sciences. Just as the dead hand of Aristotelian authority dominated pagan learning, so the ecclesiastical absolutism of the Christian Church crippled discussion of intellectual matters, substituting faith for reason as the criterion of truth. Expectation of the imminent Second Coming of Christ, as promised to his disciples, made people intellectually lazy: what was the point or need of further intellectual inquiry?

Hypatia, a neo-Platonist, had no truck with such attitudes. She also had the habit of engaging casual passers-by in dialectical explanations of philosophical questions. These character traits, and an interest in local politics, led to her death. She was caught up in a feud between Cyril, the fanatical Patriarch of Constantinople, and her ex-pupil and friend Orestes, the (pagan) Roman Prefect of Alexandria. According to the historian Socrates Scholasticus, a mob of 'cock-brains', supporters of the Patriarch, came across Hypatia during an anti-Roman demonstration in the streets. Inflamed with passion against the perceived evils of pagan philosophy, not to mention her support of the Roman oppressor, they dragged her into a church, stripped and murdered her, and then quartered and burned her corpse.

It is something of a relief to turn from this gruesome recital to Hypatia's actual work on mathematics. She wrote a number of books, intended as texts for her students. These included a commentary on Diophantus' remarks on Apollonius' *Conic Sections* and an analysis of her father's edition of Euclid's *Elements*. In addition, she invented a number of instruments – an astrolabe for astronomical observations, a distilling apparatus, a hygrometer and a water-level instrument. Unfortunately, none of these works, either writings or instruments, have survived.

In fact, Hypatia's importance for later scientists has less to do with her work than with the manner of her death. Her murder by a Christian mob was taken as symbolic by generations of European freethinkers, scientists and anti-Catholics. She was the last of the pagan scientists. Her death (in 415) coincided with the last years of the Roman Empire and the beginning of the Dark Ages. For 1000 years or so there were no significant advances, in Christian Europe, in the knowledge of nature. In 640, the city of Alexandria was invaded by the Arabs and what remained of the great library was destroyed. Scholars, alchemists and astrologers fled westwards to avoid the conquerors. They carried off many of the precious manuscripts, which were subsequently lost forever. The Arabs rescued what was left. Under the enlightened leadership of some of the Caliphs, they carried out a massive programme of translation and development of Western science. While Christian Europe sank into intellectual chaos and barbarism, the Arabs, using all the available resources of Greek, Jewish and Christian scholarship, fostered a scientific renaissance.

1 2 3 4 5 6 7 8 9 10 11 12 13 14 15 16 17 18

ANCIENT ISRAEL

Our intent is to see to the affairs of Man upon which his existence depends,
that he should see the consequences of his deeds. . . . It is a moral question.
Moses ben Maimon (*Maimonides*)

FROM ancient times the Jewish people was convinced that it was a race
apart. Jewish scribes wrote about their nation as unique, as having been
chosen by the Creator for some special purpose. The god of the Jews, the
Creator Jahweh, had a personal interest in them, having prepared an
exclusive destiny for their nation. They believed in their religion as the only
true one. All other nations worshipped idols and had many gods. This did
not offend Jahweh. His favours were reserved for the children of Israel, at
least those who carried out His commandments. The everyday life of the
Jews was regulated by sacred texts which they held to be divinely inspired.
It has been estimated that an orthodox believer, in his or her daily life, had
to avoid transgressing more than 600 rules and patterns of behaviour.

Unlike their near neighbours and ethnic cousins the Arabs, certain
sections of whom were intellectually tolerant, inquisitive and assimilative,
the Jewish people described themselves as 'stiff-necked'. Their scholars
borrowed little in the way of secular knowledge from the nations to which
they were subject at different times – whether Greeks, Romans, Egyptians
or Babylonians. They saw themselves as engaged in a religious war of
survival in which slavery, persecution and attempted genocide were their
portion, allotted to them by Jahweh. They defined their nationhood in
terms of their sacred book, the Bible. The religious observances it laid down
guaranteed national as well as individual survival. During the Egyptian
and Babylonian captivities, and even after their conquest by Rome, the
Jewish people refused all compromise in matters of religion – and since all
life, all thought, was bound up with Jahweh, the same kind of introversion
marks all their scholarship, including mathematics. It is impossible to
understand the nature of the Jewish contribution to science without

realising the extent – which cannot be overstated – to which it was moulded and shaped by religious belief and practice.

Hebrew numerals

The ancient Hebrews, like certain other nations, used their alphabet not only as letters but also as numerals – and this dual function was a serious handicap to their progress in arithmetic and algebra. Like other Semitic alphabets, theirs had 22 letters only (no vowels). They used all of them as numerals and added five more (variants of letters which take on a different form at the end of a word).

Lacking the concept of place value, and with no sign for zero, Jewish scholars had serious problems with numbers. Addition and subtraction were relatively simple, but multiplication, division and fractions were so difficult with alphabetic numerals as to be almost impossible. As a result, the Jews, like the Greeks (who had the same problems for the same reason), found it more congenial to talk about numbers in a magical and religious context than to use them as a tool for solving practical everyday problems. They distinguished between 'figures' (geometric patterns of pebbles, or scratch marks in the sand, used as aids in counting) and 'numbers' (in the form of letters). The ancient Israelites developed fantasies and alleged hidden meanings of particular numbers (they specialised in number magic and *gematria*, which assigned numbers to names as aids to prediction). The problem of numbers was further obscured by the Jewish belief (similar to that of Pythagoras and his followers in ancient Greece: see page 79) that the letter-numbers had been used by the Creator as building blocks of the universe, and were therefore sacred.

The Jewish emphasis on religion as the basis of all life led their scholars to devote themselves almost entirely to the study of Divine Law as set out in the revealed texts. Instead of engaging in elevated talk about the Deity, rabbis spent their time in systematic, logical, and pedestrian analysis of the sacred texts. This analysis was fundamental to discussion of specific problems of conduct. The discussions, recorded in the *Talmud*, run to about 20,000 pages in English translation. They span a period of almost a thousand years and record arguments by scholars of the Law from a dozen countries.

The *Talmud* is concerned, above all, with justice, both civil and religious. Many of the problems it discusses have some number aspect which calls for new ways of thinking. But these problems arise in the context of legal argument, and that argument is the centre of interest. The *Talmud* was a report by experts on the Law, a kind of research journal in which the rabbis argued their findings.

The rabbis opposed secular science, regarding it as a pagan invention. As such, it was forbidden on the same principle as forbidden food (which pollutes other foods it touches). Number and science were neglected, and ignored, to an amazing extent. Some historians even ask how the Jewish nation survived for 15 or 16 centuries in spite of having no system of written numerals: the first Hebrew numerals known appear on coins of the Hasmonean dynasty in the 2nd century. The answer is similar to what happened in Christian Europe in the Dark Ages, when Greek and Roman letters were the only 'official' numerals. Merchants, farmers and soldiers pursued their practical affairs using fingers and counters as tallies, communicating the totals orally. Tally sticks were used as demand notes and receipts in business and even by governments. If written calculations were required, Babylonian, Egyptian or Aramaic numerals and counting methods might be used. Discussion of the learned hardly impinged on such work, except in court cases – and then the experts handed down their opinions in abstract, descriptive terms, leaving it to those concerned with such matters to work out the practical results of their conclusions. It was a literate society, but (especially in the higher, more abstract areas of intellectual reasoning) scarcely numerate.

Symbol	Name	Value	Symbol	Name	Value
כ	aleph	1	כ	xaf	20
ב or ב	bet	2	ל	lamed	30
ג	gimmel	3	מ	mem	40
ד	dalet	4	נ	noon	50
ה	hay	5	ס	samech	60
ו	vav	6	ע	ayin	70
ז	zayin	7	פ	pay	—
ח	xat	8	פ	fay	80
ט	tat	9	צ	tzadeek	90
י	yood	10	ק	koof	100
כ	kaf	—			

Hebrew numerals/letters

The 'Book of Creation'

The Book of Creation (*Sefer Yetsirah*) is a commentary on the Book of Genesis. It is not part of revelation as accepted by the Jews. The Book is a short treatise devoted to the theme that numbers are the 'primaeval stuff' from which God created the universe.

God drew them, hewed them, combined them, weighed them, interchanged them, and through them produced all of creation and everything designed to be created.
(*Safer Yetsirah*, chapter 2, para. 2)

Seven of these letters correspond to the Roman b, g, d, k, p, r and t. They marked the seven days of Creation. With each of these letters in turn, God created the seven days of the week, seven planets and seven orifices of the human body, as follows:

First day: Beth, ruler of life; Saturn; the right eye. (The first day of Creation, in passing, was neither Saturday nor Sunday: it was Tuesday.)
Second day: Gimel; Jupiter; the left eye.
Third day: Daleth; Mars; the right ear.
Fourth day: Kaph; the Sun; the left ear.
Fifth day: Pe; Venus; the right nostril.
Sixth day: Resh; Mercury; the left nostril.
Seventh day: Taw; the Moon; the mouth.

As well as seven planets, seven days of the week and seven 'gates' in the body, seven heavens were created after the same pattern. In this way, the divine preference for the number seven was expressed – not to mention the role of the seven letters (that is, numbers) as the essential starting-points of creation. The order in which the planets were created coincided with the order of their accepted distances from the Earth.

The Book of Creation also says that the Sun, Moon, stars and planets play a major role in deciding human destiny. The planets gave their names not only to the days of the week and the months of the year, but also to the hours of the day. It was a feature of Jewish astrology (as of Greek) that it was the hour of birth, not the day or the zodiacal 'house', which controlled one's destiny.

The origin of multiplication

There are few Hebrew records of number import until the 9th century AD, when the Jewish people came under the influence of the Arab scientific renaissance. The chief hint of what mathematical thinking there was before this comes from certain words and phrases in the Hebrew language:

mathematical ideas, as it were, embalmed. These expressions are: *manah mispar be-mispar* (to count one number by means of another), *manah sela' be-sela'* (to count one side by another), *kafal* (to fold or double), *saraf* (to twist or weave), *hikkah* (to strike) and *arak* (to set up a proportion). They point to a trial-and-error process in thinking about number. Everyday words like 'twist', 'double', 'count' and 'fold' take on a technical meaning as well as their more usual sense. They define a situation, for example, where a certain number (the *mukkeh* or 'stricken one') is added to itself, again and again. The number of times the amount is added is given by another number, the 'striker'.

Lacking any knowledge of the sequence of linguistic events, we can only speculate about the exact number processes at work here. Even so, it is possible to arrange the operations with numbers in a sequence, from simple to complex. (This is done for convenience, not to suggest that the chosen sequence is either inevitable or historical.)

'Folding together' of many equal addenda is a simple way to think of multiplication. It grows out of the most basic of arithmetical processes, simple addition. This is *manah mispar be-mispar*, that is, counting one number (say 23) by another (say 17). You 'fold' the number 23 over on itself 17 times.

manah mispar be-mispar	*manah sela' be-sela'*	*kefal*	
1. $23 \Rightarrow 23$	Plaited mat 17	1	23
2. 23 46	units long by	2	46
3. 23 69	23 units wide	4	92
4. 23 92		8	184
5. 23 115		16	368
6. 23 138			
7. 23 161		$17 = 16 + 1$	
		17×23	
$7 \times 23 = 161$		$= (16 \times 23)$	
$10 \times 23 = 230$		$+ (1 \times 23)$	
		$= 368 + 23$	
$17 \times 23 = 391$		$= 391$	
	391 units of area		
Repeated addition method	Counting method	Egyptian, or binomial method	

The second method recognises that any number can be made up by adding selected terms from the series 1, 2, 4, 8, 16 . . . This fact was known to the ancient Egyptians, and indeed was the basis of their arithmetic. Jewish mathematicians learned it from them, and called it *kafal* (duplication, or doubling).

The third method represents a higher level of understanding of multiplication. The word *saraf* (to twist, plait or weave) is used of the plaiting of reeds or willow twigs to make a shelter. It also describes how a line is converted into an area, as in weaving a line of papyrus leaves into a sheet. It is known as *mana sela' be-sela'*, that is, counting the side by the side. The table on p. 97 shows, in modern notation, how each of these methods can be used to multiply 13 by 17.

Jewish fractions

Three kinds of fractions were known in ancient Israel. First, and least often used, were 'common fractions' or 'partitions', of the sort known today, with both numerators and denominators: $\frac{3}{5}, \frac{4}{9}, \frac{10}{17}$ and so on. Second were 'unit fractions', learned from the Egyptians. These had no numerators: they were 'unit' fractions because each consisted of one unit only, divided into parts by another number, as say: $\frac{1}{2}, \frac{1}{10}, \frac{1}{38}$ and so on. All partitions were written as a sequence of unit fractions. For example (using Arabic numbers for convenience), $\frac{2}{7}$ would be written as $\overline{4}, \overline{28}$, that is $\frac{1}{4} + \frac{1}{28}(=\frac{2}{7})$.

The third kind of fractions, sexagesimal numbers, were learned from the Babylonians, and used the Babylonian system of counting to base 60. They were written in the form (for example) 2, units; 32, 29. (This meant 2 plus $\frac{32}{60}$ plus $\frac{29}{3600}$.)

A whole section of the Rhind papyrus (a main source for Egyptian maths: see page 40) is taken up by a table of fractions of the form $\frac{2}{n}$, where n assumes a sequence of values given by the odd numbers from 3 to 101. The table shows how these are each transformed into sums of unit fractions (as in the case of $\frac{2}{7}$ above). A special hieroglyphic sign is used for $\frac{2}{3}$: ⼌ . This fraction, which has a unique place in the Egyptian system of arithmetic, can be translated as 'two mouths' – and it has exactly the same meaning in Hebrew. The *Mishna* and *Talmud* speak also of 'two parts', meaning two-thirds, and 'three parts', meaning three-quarters. The *Talmud* says, for example, that the duration of twilight is 'three parts of a mile'. This is an elliptical way of saying three-quarters of 18 minutes, the time required slowly to walk a mile. From the point of view of mathematical calculation, it is hardly the clearest or most manageable way of indicating $13\frac{1}{2}$ minutes.

Jewish logic and scientific method

All the main historical religions – Judaism, Christianity, Buddhism, Hinduism and Islam – set up the kind of discussion known as scholastic philosophy: the sustained discussion and reasoned analysis of religion over a period of many centuries. Believers organised themselves into stable groups, which developed a tradition directed to the analysis of their religious beliefs, and of 'revealed' texts, by the use of reason. (There was, for such scholars, no conflict between faith and reason.)

Jewish scholastics devoted themselves, above all else, to 'figuring out' the plan of creation, using all the resources at their disposal: sacred books, oral traditions, their thoughts about the Deity and the method of dialectical discussions and rational enquiry. They subjected the texts, the books of the Old Testament, to meticulous analysis, letter by letter, word by word. This was only fitting as the content was thought to be the word of God which had to be squeezed dry of all possible meanings.

Given such scriptural fundamentalism, it is at first surprising to find the view, stated in the *Talmud*, that 'A sage is superior to a prophet'. In other words, in matters of religion, the claims of reason, provided that it was soundly based, were of more value, and should be acted upon, in preference to the more enthusiastic but maybe extreme reactions of 'holy men' (that is, priests and prophets). Even miracles were suspect. They had to be analysed and tested for meaning and credibility. In the last analysis, they might be less weighty than the findings of sages who had devoted their whole lives to deciding questions of authority and morality.

This attitude was based on the concept of *Sebhara* ('sound judgement', or, very nearly, 'common sense') to be found in the analyses and conclusions of the true scholar. Dealing as it does in self-evident truths, it is not easily to be disputed. *Sebhara*, in fact, has the same status as logic or revelation. It is necessary to correct over-refined abstraction by means of truth-in-context. For example, *Sebhara* would argue against the ethical neutrality of science (which leads practitioners to sanction taking problems out of all human context). It would refuse to sanction experiments that work against morality and the survival of the human species. The story of the Garden of Eden (whether true or false) has two messages for the scientist. The first is that the Earth is the home of the human family; the second is that it is not to be sacrificed to satisfy some momentary, primaeval impulse.

Jewish scholasticism was first to declare the principle of economy (later known as 'Occam's razor': see page 14). This is one of the main guides used by the scientist to choose between alternative views. It asserts that there is no special virtue in elaborate and complex explanations, as opposed to a small, simple hypothesis that covers the essential facts. If it fits the general pattern of understanding and leads to new knowledge, so much the better.

The principle of economy appears in the *Talmud* in the statement that there is nothing superfluous in Scripture. The *Torah* constitutes a unity. It is not a miscellaneous collection of disparate items. This means that a general principle only needs one specific case to illuminate it. It may be that two cases are given in illustration. We then have to ask the question: Why two, when one would be enough? The answer is that two cases are given to indicate that the principle in question does not apply widely but only in these special cases.

Another unique and more direct principle is what is referred to as the *qal-wa-Homer*. The logical argument known as the *ha-kol* formula corresponds to Aristotle's syllogism, which Jewish rabbis also discovered. The *qal-wa-Homer* has also been identified with Aristotle's *a fortiori* method of reasoning. It is a logical proof that proceeds from a minor connection to a major conclusion. The Hebrew words (thought to mean 'the light and the heavy') point to the fact that it is a form of reasoning by analogy.

An example from the *Talmud* is the argument against a man accused of assaulting his neighbour. It was known that the man had once beaten his father and mother, indicating that he was a person of violent temper. The *qal-wa-Homer* declares that if he did not draw back from injuring his own father and mother (a capital offence in Jewish law), it would hardly be surprising if he assaulted strangers. Of course, direct evidence of the assault was also necessary to convict him. The *qal-wa-Homer* can be recognised because it is cast in the form: 'If such-and-such is true (and you cannot doubt it), then how much more [certain] is it that so-and-so is true'.

Another innovation made in the course of Talmudic discussion was the method of reasoning known as *binyan 'av* ('making a father'). This is not the same as reasoning by analogy; it is similar to 'the method of agreement' used in scientific reasoning to identify causes and effects.

'Making a father' starts by asking how we would justify grouping a number of things together. We first decide what should be included in the group as a preliminary to establishing the maximum of instances to be explained. None of these instances can be reduced to the others. For example, we discover a rationale to explain the various ways in which one can contract a legal marriage. The *Torah* gives four ways: (i) by a money payment; (ii) by delivering a bond or deed of marriage into the woman's hand; (iii) by the man and woman cohabiting, having declared their intention of marriage; (iv) by the man and woman coming together under the canopy or sunshade (part of the ceremony in a 'normal' marriage).

None of these ways can be derived from the others. The question is, can we recognise any common factor that justifies grouping them together as valid means of effecting a marriage? In other words, if their differences are ignored, what do they have in common?

The answer, that 'the power to marry does not come from betrothal but is acquired elsewhere', is the 'father', created by a systematic testing of all alternative 'fathers', none of which fits all cases. The final stage of the argument can be set out as follows:

(i) by a payment of money (*keseph*). A payment can be made in place of carrying out a promise to consecrate gifts to be sacrificed in the Temple; so a money payment can take the place of a promise to marry.

(ii) delivery of a bond into the woman's hand (*shetar*). The binding element in a betrothal lies in the exchange of contracts to marry; therefore a bond can substitute for the group ceremony.

(iii) intercourse with a commitment to marry (*bi'ah*). Since such intercourse is all that is necessary in a levirate marriage (one in which a man marries his dead brother's widow), it should also be as acceptable as a properly witnessed betrothal. No actual ceremony is needed; it has the same legal status as our common-law marriage.

(iv) the man and woman going together under the canopy (*hullah*). This occurs at one stage of the 'normal' marriage ceremony as symbolic of the desire to share each other's life-space in perpetuity; it can act in the same way, retrospectively, as a substitute for the betrothal.

The method is similar to John Stuart Mill's description of how the scientist sets out (if only in his or her own mind) the circumstances in which a particular phenomenon occurs and when it does not. The 'method of agreement' states that, when there is an invariable association between certain events which happen before the phenomenon, and an absence of the phenomenon when these events do not occur, then the events are related in a causal manner to the phenomenon.

To illustrate how similar Mill's analysis is to that of the *Talmud*, we can follow the lead given by Robert Boyle and check what causes air and other gases to expand and contract. We set up a number of experiments. Using the appropriate apparatus, we trap some air in an enclosed space. Then we test various factors to discover what changes might affect its volume. We discover that only two things seem to matter: the pressure that we impose on the gas (low or high) and the temperatures to which we subject it. High pressures cause the air to contract; it expands when heated. Nothing else seems to affect the volume it occupies. So we say that 'the volume and temperature are directly related, and the volume and pressure are inversely related'. This is the *binyan 'av* known as Boyle's Law.

Another important discovery enshrined in the *Talmud* is the *en la-dabhar soph* mode of proof. This is the *reductio ad absurdum* argument also attributed to such Greek thinkers as Theaetetus and Zeno. It tests some declared truth or opinion by drawing some conclusion from it which is known to be false. If this can be done, the opinion itself must be false. An example can be given from the Bible itself. In Numbers 5 : 11–31 we are told how the law of jealousy works. A husband suspects his wife of associating with another man. He takes her to the temple where the priest will test her guilt by ordeal. He prepares 'bitter water' by mixing dust from the floor of the tabernacle with holy water. He forces his wife to drink the brew while the priest utters a curse on her, to take effect only if she is guilty. In that event, the bitter water will cause her belly to swell and her thighs to rot. The priest writes the curse in a book, and presumably gives the result.

In the *Talmud* the problem centres on the question whether God might halt the effect of the curse on a guilty wife who was otherwise a woman of great merit. This argument was demolished by showing the absurdity of such a view. Such arbitrary behaviour on God's part would put in question the results of all such tests. If 'bitter water' was known not to work even in one case, the issue would always be in doubt as to whether a guilty woman was being shown special favour by God. But the good name of the innocent would also suffer. There would always be a question as to whether the innocent woman, too, was guilty but being shown special favour. Perhaps more important, some guilty women would escape just retribution. This would call in question God's justice. The original suggestion is hereby reduced to an absurdity.

The *Talmud* parts company with the accepted structure of argument in science, in recognising two kinds of laws. The first is *hoq* (a law which contradicts reason). As an example, we have the dietary laws. These are not irrational, but could not have been discovered by experiment or logial analysis. They have simply been declared. In the second case (*hiddush*), the laws cannot be explained by reason nor can they be used to justify new laws. For example, there are a number of anomalies in the law of false witness. The false witness is disqualified from testimony – but only from the time it is discovered that his (or her) evidence is false. His evidence as a whole is not excluded. If the crime is capital, he is punished by death – but only if he has given false witness about himself (for example, where he was when the crime was committed). If his false evidence relates only to the accused, then he is still punished but not to the extent of the death penalty. These anomalies cannot be explained in terms of rational analysis.

It is clear from all this that Jewish interpreters of the laws could not be accused of credulity, nor was their mode of reasoning about events in any way contrary to the principles of inductive science. In fact, they anticipated many of the basic laws of statistical reasoning. Extensive use of the method of casting lots and constant scrutiny of the system for fairness (because of the tendency of individual priests to manipulate the system for personal gain) was a strong motivation to study the operations of chance. Expressing the idea of the random principle led to the recognition that a number of repetitions of the process of choosing should result in an equal number of choices of the alternatives. The fact that, in the short run, there might be a bias in the results of the draw but that this bias would disappear in the long run of repeated samplings, led them to an intuitive awareness of the law of great numbers (that equal probabilities yield equal frequencies as the draw is repeated more and more times). The recognition that 'counted majorities' yield numbers equal to the actual probabilities which operate in cases of unbiased sampling was also recognised. The need to repeat the lottery several times if one wanted accurately to determine the will of God anticipated a good deal of modern experimental method.

Select cases

The survival of the Jewish people is a direct result of their belief in their destiny as God's chosen people. According to the Scriptures, Jahweh made a covenant with Moses to lead His people from slavery into the Promised Land. Here their nation would survive forever, on condition that they preserved God's Law intact. This meant that they must forswear all other gods, make sacrifice to Him, as laid down in revelation, and carry out the rituals that He demanded.

Upholding the Law involved not only daily prayers and rituals, but also abstaining from certain foods. Some were forbidden because an animal was ritually unclean (like the pig), or was diseased or malformed, or had not been killed in accordance with due ritual. In other cases, the food (animal, fruit or grain) might be forbidden because it was reserved as an offering for priestly consumption, or forbidden even to priests because it was God's portion. Forbidden food could come into contact with, and contaminate, other food, bringing it under the same prohibition. Some rules of avoidance were absolute, as in the case of swine's flesh. In other cases, 'forbidden' foods became available on the basis of rules which narrowly defined the degree of contamination permitted. Of these food proscriptions, probably the best known is that meat not ritually killed and declared *kosher* by a rabbi is forbidden.

Kosher and non-kosher

The *Talmud* examines a number of situations involving *kosher* and non-*kosher* meat. These are cases where a decision must be made as to whether a piece of meat, already eaten or about to be eaten, is forbidden. In the first case, the facts are given as follows. In a certain village, there are 10 shops that sell meat. Nine are *kosher*, the tenth is owned by an Arab whose meat is non-*kosher*. A piece of meat is found in the village. The question is, can it be eaten or is it prohibited? Since there is no information about where the meat comes from, the court must decide on the basis of what other facts are known. It decides that since the majority of the village shops are *kosher*, the balance of probability is that this particular piece of meat can be eaten. The court 'follows the majority' and declares the meat *kosher*.

This case differs from that of a person who buys meat in a village shop, without knowing whether it is *kosher* or not. Here the question is simply, is the meat *kosher*? There are two possibilities; this means that the chances are half and half. In the absence of other evidence, the question must be left unanswered. Therefore we must give the benefit of the doubt to the suppliant and accept the food as *kosher*. This decision, however, is merely legal, allowing the meat to be eaten without penalty. It leaves open the question of fact, whether the animal was actually *kosher* (that is, that all rituals were properly carried out). Some might argue that a better decision might be to defer the ruling while fresh evidence is sought: the ratio of *kosher* to non-*kosher* shops, for example. It is in no way a contradiction that we seem to get two different answers to the question of probability: 1 in 10, and half and half. They are answers to two different questions. In either event, the decision is the same, that the meat is *kosher* and may be eaten.

In the *Talmud* discussion of the next case, there is what looks like a piece of specious reasoning. A parcel containing 10 portions of meat has been found in the same village. It is known that there is one piece from each of the butcher's shops in the parcel. In other words, nine pieces of meat are *kosher* and one is not. How do we advise the finder?

The new principle here, used by Rabbi Schlomo ben Adret to answer this question, is 'the majority nullifies'. The rabbi says that he would advise the finder to choose any piece of meat and would then say 'This is not the one. Eat.' The rationale is that the chances are nine to one that the piece is *kosher*. He would repeat this statement for each piece until there were only three left. At this point he would still say, 'This is not the one, there is a majority. Eat.' With two pieces to choose from, he would say, 'Eat. The prohibited piece is (in your stomach) with the majority.' The same formula would be used with the last piece.

The reasoning here is expressed in a general form as 'one in two is nullified': that is, when one portion of non-*kosher* meat is mixed with at least two portions of *kosher* meat, and cannot be identified, all three portions may be eaten. This allows us to eat the first piece of the 10 even though it may be non-*kosher*. The same applies to pieces nine to three: in each case, pieces of *kosher* meat are in a majority. When we have two pieces, we are allowed to eat the penultimate one since we can concede at this point that it is much more probable that the previously forbidden piece has already been eaten. The last piece has at least a half-and-half chance of being *kosher*; the eater is given the benefit of the doubt. (Strictly speaking, the chances that the last piece is, in fact, *kosher* are not half-and-half, but 9-to-1. It almost certainly is *kosher*.)

This line of argument may appear to a non-believer to be a monstrous piece of casuistry. But our interest is in understanding the logic, not in disagreeing with it. The point is, that the decision must be made according to rule. 'Eat, it is not forbidden' is a legal declaration, not a statement that the piece of meat is ritually pure. The situation is exactly analogous to the rules of evidence in a modern court. Although a criminal may be manifestly guilty as charged, the court must declare him innocent if (say) a confession has been extorted from him before advising him of his legal rights, or by torture.

In all the above cases, the majority can be stated as a number, the amounts (of pieces of meat or shops) are known. In other cases, the majority has not been counted, but is known to be substantial. For example, we can say that the great majority of children will be fertile when mature. There is no way we could prove this except as a hypothetical extrapolation from previous generations. However, the question is not academic, but concerns an issue vital to many. In the case of levirate marriage, the moral justification of enabling, or even compelling, a deceased husband's brother to marry the widow is that by this means the deceased is enabled to procreate by means of an acceptable surrogate. Therefore, if the surviving brother is known to be infertile, the law cannot be enforced. It is easy to imagine circumstances where this issue might come to court.

The idea of an uncounted majority is important in many other situations. For example, in most orthodox Jewish villages (the Polish *stetl*, for example), it would hardly be necessary to count the number of *kosher* and non-*kosher* shops. Non-*kosher* shops would be in a tiny minority, or non-existent. In other food prohibitions, by contrast, exact quantities must be known. The question of the actual number in the majority is of central importance.

Fruits of the Earth

As we have seen, 'one in two nullifies' in the case of *kosher* meat. With respect to grain and fruits, there are a number of prohibitions. There is the obligation of *terumah* (the tithe), whereby an offering is set aside from the grain crop, to be eaten by priests of the temple under conditions of ritual purity. It is forbidden to others unless it happens to have been accidentally mixed with other grain. Even then, the proportions must be right: at least 100 parts of normal grain to one of *terumah* grain. As in all such cases, there is no way we can decide from appearance alone which foods are *terumah* or permissible. The same prohibition, and the same qualification, applies to eating *hallah*, a temple offering of grain in the form of dough, and to first fruits. Similarly, if a tree bears fruits before the third year, this fruit is known as *orlah* and is forbidden for all purposes, even to priests. The prohibition is waived if the proportion of *orlah* to ordinary fruit is less than 1 in 200.

In all food prohibitions, of which these are examples, ways are found to avoid absolute prohibition. (The one exception is contamination by snake venom, in which the principle that 'danger is greater than prohibition' is applied, and there is no relaxation of the rule.) Each way gives rise to lively Talmudic discussion. The first way, as described above, is to allow permitted food to nullify the forbidden food when the ratio passes a certain level. Even *orlah* is accepted if the ratio of normal fruit to *orlah* is 200 to 1 or greater. The second way is to allow one prohibited food to neutralise another. For example, it is written, '*Terumah* can neutralise *orlah*'. Examples to show how these (not uncommon) cases were decided can be given in a table as follows:

---

100 *seah* (dry units) of ordinary fruit
  1 *seah*   of *terumah* fruit

Total 101 *seah*; ratio 100:1
The mixture can be eaten.

---

202 *kahs* (half units) of permitted fruit
  1 *kah*   of *orlah* fruit

Total 203 *kahs*. Ratio greater than 200:1
The fruit may all be eaten.

In each of these examples, the *terumah* food is first legitimised and then added to the *orlah* to form a permitted mixture. Counting the whole mixture establishes an acceptable mixture where the *orlah* is neutralised in part by the *terumah*.

The woman accused of adultery

This suit was started by a man who claimed that his wife was not a virgin at marriage. He was anxious for a divorce without penalty, that is, without having to pay back the dowry. She confessed that she was not a virgin at marriage. It would have been difficult for her not to do so, as the sheets from the marital bed had been exposed to the wedding guests the morning after, as was the custom. Several witnesses were prepared to testify that the test for virginity was negative. The two questions which the court had to decide were (i) Did the illicit act take place 'under him', that is during the period of the betrothal? (ii) Was it with her consent or against her will? On the basis of the answers, it would be decided whether she was guilty of adultery and liable to punishment.

There was no direct evidence relevant to either question. This meant that there was a double doubt. The court's decision could only be based on probabilities. This made it difficult because the rabbis had adopted the (correct) theory that the probability of events was based on relative frequencies. The problem was that there was no way of knowing how many women commit adultery during the betrothal period and how many after marriage. In the absence of knowledge whether it happened 'under him' or not, the probability had to be assumed to be one-half. (This is not a guess at the answer, but simply records that one answer is as good, or as bad, as the other when we have no certain knowledge.)

The same reasoning applies to the question, Was it rape (answer, probability one-half); or was it with consent (answer, probability also one-half)? The probability that the illicit intercourse was both 'under him' and with consent is then only one in four (half times a half). This is a minority: the decision is therefore that the woman is not guilty.

This decision left plenty of room for argument. The first point is that even if we add all the cases of 'rape' to cases of 'not under him', it cannot be ruled with certainty that the woman was not guilty. This, of course, is true, but it merely states a truism. All we have to do is to establish whether there is a doubt. If so, she must be given the benefit. In the absence of evidence this is the only sound procedure. If there is a doubt, the accused is entitled to a ruling in her favour.

The doubt in some minds was whether the betrothal took place when the accused was still very young, and that she may have been seduced whilst still a minor. Seduction of a juvenile was classed as rape in the Jewish law. Some might argue against this, that 'the category of violence is one': a legal aphorism pointing out that the question of rape or consent is simple and should not be confused by making subtle distinctions. By classifying acts as

> violent rape
> seduction of a minor
> rape but with tacit consent
> seduction of an adult
> intercourse as a result of female aggressiveness

we merely confuse the issue, whereas if we set out the double doubt, only two possible answers to each question are possible:

> It happened:
> A = after betrothal / not-A = before betrothal
> V = with violence / not-V = with consent
> There are then only four possibilities to consider:
>
> | V/A | V/not-A | not-V/not-A | A/not-V |
> |---|---|---|---|
> | innocent | innocent | innocent | guilty |

With this approach, the indications are still only one in four that the woman is culpable. She must therefore be acquitted of adultery.

There remains a different kind of objection of some theoretical significance. The categories 'under him' and 'willingly' are quite independent, and so each should stand alone. Therefore, they should be opposed separately to 'under him willingly'. That is, the half 'under him' is in reality opposed to the 'not under him' whereas the half that counts as 'rape' should be opposed to 'willingly'. Thus the category of 'rape' opposes 'willingly', annulling it. This leaves only the even doubt whether it was 'under him' or not. The split is therefore only half-and-half, not a quarter-by-three-quarters. The court decision based on this probability is more open to doubt, but it does not change.

The widow remarries: whose child is it?

A certain widow, exercising her right under Jewish law, married her brother-in-law on the death of her husband. He was taking advantage of the law of levirate, designed to ensure that a deceased male leaves a family to carry on his name. There was no need for a ceremony. The fact of cohabitation by the widow and a close male relative of the deceased, with the intent to marry, was enough to establish the marriage as a fact. The woman had only to wait three months from the death of her husband.

In this case, there was no sign that the widow was pregnant before her second marriage. But six months later she gave birth to a child. The question to be decided was, who was the father? To put it otherwise, was the child a full-term baby (the first husband being the father), or was it premature (the second husband being the father)? The question was relevant to the child's inheritance rights. If the mother was already pregnant by the first husband when she married his brother, this would affect the child's rights to inherit from the brother. It would invalidate the marriage, and there would be complications about the woman's right to support. A good deal hinged on the court decision.

In deciding such cases, the court was guided by certain legal presumptions (known as *hazakah*). These were axioms or principles which all parties agreed to, unless there was clear evidence that they did not apply. (In American law, they are somewhat similar to 'stipulations', where the lawyer, usually for the defence, declares in the opening address that certain issues are accepted, without any proof being asked for.)

In the *Talmud, hazakah* may take a form such as, 'A process once set in motion will run its course unless something happens to stop it'. In this case, the *hazakah* would suggest that (i) a pregnancy once started continues until the child is born, normally a period of nine months after conception; (ii) with most women, pregnancy is apparent three months after conception.

In this case, these assertions are matters of dispute. There is no question here of an immediate ruling, based on deductive logic, that the first husband is the father because events can be taken to follow the normal laws of nature. There are enough exceptions to the nine-month and three-month rules to raise doubts about the child's paternity. Perhaps, for example, the pregnancy did not 'follow the majority', the decision rule in cases involving probability. In any event, the problem was not to be solved by deduction from declared axioms, with the conclusion certain. Everyday problems respond to a different logic, based on induction and inference. In this logic, a conclusion can be reached only on the basis of what events are most likely. In other words, in the absence of conclusive evidence, a decision between two hypotheses can be based only on probability.

As we have seen, nine *kosher* shops in ten was by *Talmud* standards a large majority. The reverse ratio, of one in ten non-*kosher* shops, counted as a small minority. Other categories of majorities and minorities, recognised by the Talmudic thinkers, are shown below.

Numbered amount	Conclusion
9 out of 10	Very large majority; quite likely
7 out of 10	Large, significant majority; likely
5 out of 10	Half-and-half; a doubtful case
3 out of 10	Small minority; unlikely but possible
1 out of 10	Very small minority; quite unlikely

In the case in question there are two alternatives: (a) the child is full-term, the deceased husband being the father; or (b) the child is premature, the brother being the father. If we suppose (a) to be true, then we need to know that only a minority of pregnancies (say two in ten) do not show after three months. Therefore, if the deceased husband was the father, the widow's pregnancy would be very likely to show when she married the brother. The chances that the pregnancy would not be detected at three months would be more remote. It could happen, but only in a minority of cases (in fact, in two cases in ten).

The second half of the argument is about whether the child is premature. The answer is clear, but not certain. Premature delivery happens in a minority of births. The chances of this baby being premature are therefore small – in fact, about one in ten. In the majority of cases, that is, about nine times in ten, babies were likely to be full-term.

Probability

The factor that ensured the primacy of Jewish thinkers in pioneering ways of thinking about chance in practical situations was that Orthodox Jews were forbidden to gamble. The Greeks might have discovered probability, for they were addicted to dice-throwing. But the dice they used were made from the astragalus bones of a sheep (what would be the knucklebones if sheep had knuckles). These were very irregular in shape, so that dice were biased and there was not an equal likelihood of each side falling uppermost. Because of the astragalus' imbalance, it was not a random device, so it was impossible for the statistical law of great numbers ('If the number of tests of an event is large, then the proportion of successes in the tests is close to the likelihood of the event') to manifest itself. By contrast, the elaborate system of casting lots used by Jewish priests did exactly what it was devised to do; that is, gave everyone an equal chance of being chosen.

The historical origin of the lottery system is that annual choice of one of two goats to be the 'scapegoat', carrying the heavy load of sin into the wilderness as the Goat of Azrael. (The other goat was sacrificed to the Lord.) Since the choice here was between only two participants, it would be demonstrable that the chances of being chosen were equal for both. This idea could be easily extended to include the whole group of temple priests who were daily caught up in casting lots for all sorts of purposes. The concept of an equal chance of being chosen in a 'fair' lottery would be the prime motivation of the whole system. In turn, the notion of equal probabilities, and the fact that such chances would be shown by the equal numbers of votes for each participant over a period, would naturally emerge. Thus the statistical law of great numbers (quoted above in the words of Jacob Bernoulli, 1713, who was credited with formulating it) was actually first declared in the *Talmud* by Rabbi Aramah in the 15th century, as follows:

Ordinary lots due to chance are without any tendency to one side or the other . . . They are not a 'sign', for matters of this kind are not established unless they are found many times . . . The casting of a lot indicates primarily a reference to chance.

The question of the widow's child can be restated in the kind of language familiar to those who made use of the lottery situation as a simulation of 'real' life. Suppose we have access to a large population of babies. If things work out normally, nine out of ten will be full-term and one-tenth will be premature. Eight in ten of the mothers will have shown signs of pregnancy within three months of conception; two in ten will not. These are notional figures.

Let us write each baby's name on a piece of paper and draw one of these papers from an urn, as though drawing lots. The question now is: What are the chances that the child whose name is drawn by lot will be (i) three months premature (that is, his mother only becoming pregnant after her second marriage), or (ii) a full-term child (whose mother was three months pregnant by her first husband but gave no indication of her condition on contracting the levirate marriage)?

Consider first the possibility that the deceased husband was the father. This would be a majority situation, in that it would be a full-term baby (like nine-tenths of births). But it would be only a small minority of this majority, in that the woman showed no signs of pregnancy at three months (like the minority of two in ten of pregnant women). To deal with both conditions at once, we must multiply the two probabilities. This gives us an estimate of the effects of the double action.

In ancient Israel, this would come out as a large majority (almost a certainty) multiplied by a moderate minority. In modern parlance, it would be the product of nine out of ten by two out of ten. In other words, it would be likely to happen 18 times in 100 pregnancies ($\frac{9}{10} \times \frac{2}{10} = \frac{18}{100}$).

This can be compared with the situation where the second husband (the brother-in-law) is taken to be the father. The chances of this being true would be a minority (since only one-tenth of births are premature). However, we can be certain that the widow would show no signs of pregnancy on re-marriage. (The probability here is 10 out of 10: a certainty.) In other words, if the brother-in-law was the father, we can be sure that his new wife was not pregnant. A levirate marriage is invalid if the two parties have intercourse within three months of the first husband's death. They are also punished for breach of the law.

In the case in question, the judges rejected the possibility that there had been such a breach. A levirate marriage is legal as soon as it becomes public knowledge that intercourse has taken place. But this must be at least three months after the death of the first husband. The delay avoids problems such as arise in this case. We can be pretty sure that, in a levirate marriage, the wife would announce as soon as possible that the marriage had taken place. The small minority of one-tenth (or $\frac{10}{100}$), multiplied by unity, shows that it is less likely that the brother-in-law is the father. The comparison is between a probability of 18 chances in 100 that the deceased husband is the father as against only 10 in 100. The first or deceased husband is nearly twice as likely as his brother to be the father.

In Europe, it was only in 1763 that this kind of double-barrelled question could be asked, and answered precisely. A scientific formula was worked out by Thomas Bayes. The rabbis solved the problem very much earlier, but expressed the argument in words, not numbers. They also thought of the analysis as a way to solve moral and legal problems, not as an end in itself. The degree of precision that they aimed at was quite adequate for this.

Working with whole numbers, we can verify the rabbinical analysis by means of Bayes' theorem. His method was as follows: work out the relative probabilities (in this case 18 and 10), add these to give the total probability, 28; then divide each relative probability by this total. Using his theorem, we can confirm that the court decision was correct. The chances that the deceased husband was the father are 9 in 14; that his brother was the father are 5 in 14. The relative probabilities remain the same, but we now know the absolute probabilities as well. The chances are similar in both cases (but not quite 50:50). They still favour the deceased husband as father by about $6\frac{1}{2}$ in 10, against about $3\frac{1}{3}$ in 10 in favour of his brother as the father.

There is no suggestion here that the *Talmud* discussion anticipated Bayes' theorem. It lacks the clarity of Bayes' analysis. We must remain content with relative, not absolute probabilities. None the less, the rabbis understood the logic underlying this analysis. They recognised the need for an estimate of prior and posterior probabilities, of evaluating (if only in words) the different levels of credibility of different hypotheses, and the need for some method of establishing these limits.

The Jewish contribution

In spite of their ancient weakness in mathematics (which can be traced directly to their use of alphabetic notation), the Jewish rabbis must be given credit for inventing a completely new way of thinking. In their concern to apply the scriptures to the minutest details of everyday life, they supplied not only the definitive method of interpreting the sacred text, but also a logical basis for an estimate of the probabilities needed by the courts to settle disputed cases. The theory is virtually the same as was developed in mathematical form by Cardano, de Moivre, Laplace, Gauss and Pearson. But in the religious context in which the rabbis worked, they not unnaturally avoided abstract speculation as an end in itself, and concentrated their analysis on empirical realities. They worked on problems by means of an inductive logic quite foreign to the speculative and mythopoeic urges of (say) Greek philosophers. In so doing, they helped, in the same way as Christian mediaeval theologians, with their analytic methods and dialectical disputation, to lay the theoretical foundation for the scientific revolution in Western Europe. They set out, albeit in rhetorical and not scientific terminology, the logical principles basic to the inductive science of modern statistics.

1 2 3 4 5 6 7 8 9 10 11 12 13 14 15 16 17 18

THE INDIAN LOVE-AFFAIR WITH NUMBER

Do not imagine that mathematics is hard and repulsive to common sense.
William Thomson, Lord Kelvin

FROM the time of their earliest civilisations, the inhabitants of the Indian subcontinent had a highly sophisticated awareness of numbers. For example, the people of Mohenjo Daro, an Indus Valley civilisation of some 5000 years ago (2550–1550 BC) used a simple decimal system, and had methods of counting, weighing and measuring far in advance of their contemporaries in Egypt, Babylon or Mycenaean Greece. Successive invaders (Aryans in the 16th century BC, Persians in the 6th century BC, Greeks under Alexander the Great in the 4th century BC) all contributed ideas to an enormous cultural (and not least mathematical) efflorescence which reached its first glory under the Gupta dynasty (4th century AD onwards) and was further enriched by Chinese and Arab scholars at a time when most of Europe was culturally in limbo.

From the time of the Aryans onwards, no other cultural forces equalled the impact of Hinduism on Indian society and intellectual life. Hindu religion, its sacred language Sanskrit and its caste system survived the attentions and influence of incomers from all kinds of alien cultural traditions, from Chinese Buddhists to Muslims, from Zoroastrians to English Christians. From the mathematical point of view, the most significant Hindu contribution to world knowledge was the decimal system, which they developed from the counting methods of Mohenjo Daro and perfected by three inventions. These were first, the use of special number-symbols unconnected with any 'outside' influence, such as letters of the alphabet or pictures of fingers and toes; second, the use of a place system (in which the value of a number depends on its position in the units, tens, hundreds, thousands place, and so on); and third, but most important of all – and a milestone as vital in the history of civilisation as the invention of the wheel] the use of a symbol for zero, to show that the place in

question adds nothing to the number. These discoveries were the foundation of centuries of Hindu superiority in arithmetic, algebra and trigonometry. Trigonometry was very largely a Hindu invention. An amalgam
of geometry and algebra, useful wherever lengths and angles have to be
calculated, it was used first in Hindu astronomy, and became of prime
importance in geometrical survey work and engineering drawing.

Hindu geometry and the Vedic altars

In the Rig-Vedas, the sacred books of the Hindu religion handed down
from a remote period, every male head of a family was obliged to perform,
every day, certain acts of worship known as *purvas*. For this purpose he had
to set up in his house three kinds of fire, protecting his house and the fires by
placing them in altars of special design. The fires were known as Dakshina,
Garhapatya and Ahavaneeya. The altars, intended to shield the fires, had to
be built to design plans which related them to each other in shape and area.

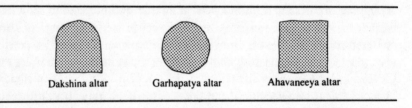

Dakshina altar   Garhapatya altar   Ahavaneeya altar

More elaborate ceremonies, for example, the offering *kamyagnis*, for the
good of the community, were performed at more elaborate altars. Some
ceremonies proceeded by stages from one special altar to another.

The altars were made of tiles, exactly constructed to sacred measure.
Their areas were related in simple ways: for example, one altar might be
twice or three times the area of the next. To calculate the correct
dimensions, sound knowledge of geometry was required. For example, the
problem might be to construct a square altar equal in area to a given
circular altar. This is quite a complex calculation, requiring an accurate
value of pi and a knowledge of the formula for the areas of a circle and
rectangle. In such ways as this, the religious concerns of Hindu heads of
families had the effect of raising the arithmetical standards of everyone
involved.

Detailed instructions of how to construct the altars were written down in
the *Sulva* Sutras, expansions of scriptural texts in the Rig-Veda (the
Samhiba, *Taittireeya Samhita* and *Taittireeya Brahmana*). *Sulva* means a

rope or cord, and was the original Sanskrit name for geometry. (By the same token, Egyptian temple surveyors are known as 'rope-stretchers'.)

There were seven *Sulva* Sutras, each named after the sage who composed it some time between 800 and 500 BC. The Sutras explain simple geometrical constructions and the theorems having to do with triangles, rectangles and circles. They do not provide a formal, systematic treatment of geometry but are simply adjuncts to religion. Nor were they built upon by later mathematicians.

The theorem about the relationship of the squares on the sides of a right-angled triangle, wrongly nowadays credited to Pythagoras, was known and widely used in ancient India. The so-called Pythagorean triples, giving the lengths of the sides of a right-angled triangle (such as (3, 4, 5), (5, 12, 13), (7, 24, 25), (8, 15, 17) and (12, 35, 37)) are found in the *Sulva* Sutras. It is possible that, like other Hindu mathematical knowledge, they may have been developed from other sources: they were also known, for example, in Babylon, Egypt and China, all of which traded with India at the time.

Unlike the Greeks, Hindu mathematicians were not at all disconcerted by the fact of incommensurability – that is, that certain numbers are never-ending and can never be exactly calculated. The existence of 'irrational' and 'absurd' numbers upset the Greeks because they thought that God created the universe from whole numbers. The Hindus, lacking any such ideas, had no cause for alarm.

As a rule, Hindu scholars wrote their mathematical problems in verse. They used pleasant and flattering words to encourage the student, made deliberately playful use of unimaginably large numbers, and obviously used maths for recreation. This love-affair with figures, so different from the awe or veneration of other peoples (for example the Greek feeling that numbers were part of a sacred mystery) has characterised Indian mathematics from earliest times.

Jaina mathematics

The Jaina religious centre at Pataliputra (modern Patna, in north-east India) had a school of mathematics for several centuries, from the 1st century BC onwards. The founder of the sect, Mahariva (6th century BC), had been a mathematician, and mathematics had always been part of Jaina religious teaching. Several holy texts dealt with *Ganitanuyoga*, or the system of calculation. It is possible that Pythagoras and Plato studied Jaina mathematical ideas, and 1000 years later Aryabhata (see page 119), the first great Indian mathematician whose name is known to us, was associated with the school.

According to Jaina cosmology, the Earth was a huge circle, divided into seven equal parts by six parallel mountain ranges running east to west. Its population was a number that could be divided by 2 ninety-six times: that is, some number of the order of 10,000 million, million, million, million.

According to the text *Sthananga Sutra* (c. 300 BC), the topics dealt with by Jaina mathematics when the school was at its height were ten in number:

Parikarma: the four operations in arithmetic
Vyavahara: concrete application to problems
Rajju: geometry
Rasi: solid geometry
Kalasavarma: fractions
Yavat-tawat (or 'unknown numbers'): algebra
Varga: squares
Varga-varga: powers and roots
Ghana: cubes
Vikalpa: permutations and combinations

Jain number theory stated that numbers were of three kinds: enumerable, non-enumerable and infinite. Enumerable numbers started at two and proceeded by units to the highest possible number. This large number can be imagined as the total number of white mustard seeds that would be needed to cover the whole Earth, filling up the oceans and valleys. Having reached it in our counting, we would carry on in stages to reach infinity. The first stage would be to continue until we reached the square of the largest enumerable number, then the fourth power, then the eighth, and so on. There were not one, but many non-enumerable numbers on the road, and once we reached infinity we would need to start all over again.

The reason for this last, extraordinary premise was that the Jains believed in five kinds of infinity. There was positive infinity, that is, infinity in one direction. There was negative infinity, reached by counting in the opposite direction, starting from minus one. There were also infinity in terms of area, and an infinity of time.

(In fact, fascination with very large numbers was an Indian characteristic in general, not confined to Jains. In the *Lalitha Visthera*, a Buddhist work written in the 1st century BC, Buddha undergoes an examination on the decimal system by the mathematician Arjuna. He is asked, among other questions, to detail the names of the numerals in steps of 100, starting with 10 to the power 7 (one *koti*, which is equal to 10 million), and running through the succeeding numbers as far as 10 to the power 53. Similarly, in the ancient Indian measurement of time, large numbers made stunning appearances. For example, one *purva* equalled 75,600,000,000,000 years, and one *Shirsha Prahalika* equalled 8,400,000 to the power 28 *purvas*.)

Lastly, it was a Jaina mathematician, Halayudha (3rd century BC), who first recognised the arithmetic triangle for which Pascal is given the credit. (It is better called the Meru-Prastera rule and was discovered about 13 centuries before Pascal.) It has to do with the coefficients of the successive terms in the expansion of the binomial expression $(a+b)^n$. It is also basic in the calculation of probabilities (see page 183).

The pulveriser

The indeterminate equation, also known in ancient China (see page 60), was a favourite study of Indian mathematicians. This is an equation where there is not just one solution but a great number. You choose the most congenial, having indicated that there are still others. Indian mathematicians from the earliest period worked on such problems, and the first Indian teacher of mathematics whose name is known worldwide, Aryabhata (475–c. 550), gave a method of solution for first-degree indeterminate equations. His method is known as the *kuttaka* or 'pulveriser'. This refers to the idea that when you work with two numbers, you batter them together until they have been ground down or pulverised. Then you put the pieces together again to yield the solution to the original problem.

The kind of real-life situation referred to in Aryabhata's discussion is similar to the indeterminate solution given in the Chinese case (see page 63). His 'pulveriser' method of solving the problem is, however, at once simpler than the Chinese and more sophisticated. An indeterminate equation enables you to find a number when you know only the remainders that are obtained when you divide by a succession of other numbers.

For example, let us set out to find an unknown number which, when divided by 137, leaves a remainder of 10; when divided by 60, leaves no remainder. These relations can be stated in the form of a first-degree equation:

$$137x + 10 = 60y$$

The pulveriser starts with the two divisors, 137 and 60. The larger is divided by the smaller, then successive divisions of the new remainders are made, as shown on the left of the table below:

```
      2*            Now form the table
 60)137
    120             *2      2      2      2.     297.
     17 (remainder) *3      3      3.     130.   130.
                    *1      1.     37.    37.
      3*            *1.     19.    19.
 17)60              18.     18.
    51              1.
     9 (remainder)

      1*
  9)17
     9
     8 (remainder)

      1*
  8)9
     8                                           x = 130
     1 (remainder)                               y = 297
```

The sources of the numbers in the right-hand column of the table above are as follows:

  (i) 2, 3, 1, 1 are, in order, the results we obtain when we divide.

 (ii) 18 is chosen as a number which, multiplied by the last remainder, 1, and subtracting 10 (extracted from the original equation), gives a number divisible by 8; i.e. the penultimate remainder must be $1 \times 18 - 10 = 8$.

(iii) these numbers

$$18 \times 1 + 1 = 19$$
$$19 \times 1 + 18 = 37$$
$$37 \times 3 + 19 = 130$$
$$130 \times 2 + 37 = 297$$

are followed by dots (as markers) in the table.

One solution of the equation is $x = 130$, $y = 297$. Other solutions can be found by adding (or subtracting) 60 for new xs and 137 for the corresponding new ys:

$$x = 10, 70, 130, 190, 250 \ldots \text{continue to add } 60$$
$$y = 23, 160, 297, 434, 571 \ldots \text{continue to add } 137$$

An infinite number of solutions can be found this way.

The *kuttaka* is a difficult process to follow. Another example may make it clearer. This uses the indeterminate equation

$$19x + 5 = 12y$$

$$\begin{array}{r} 1* \\ \hline 12\overline{)19} \\ 12 \\ \hline 7 \text{ (remainder)} \end{array}$$

$$\begin{array}{r} 1* \\ \hline 7\overline{)12} \\ 7 \\ \hline 5 \text{ (remainder)} \end{array}$$

$$\begin{array}{r} 1* \\ \hline 5\overline{)7} \\ 5 \\ \hline 2 \text{ (remainder)} \end{array}$$

$$\begin{array}{r} 2* \\ \hline 2\overline{)5} \\ 4 \\ \hline 1 \text{ (remainder)} \end{array}$$

Now form the table

*1	1	1	1.	59.
*1	1	1.	37.	37.
*1	1.	22.	22.	
*2.	15.	15.		
7.	7.			
1.				

$$x = 37$$
$$y = 59$$

The sources of the numbers in the right-hand column of this table are as follows:

(i) 1, 1, 1, 2 are, in order, the results we obtain when we divide.
(ii) 7 is chosen as a number which, multiplied by the last remainder, 1, and subtracting 5 (extracted from the original equation), gives a number divisible by 2; i.e. the penultimate remainder must be $1 \times 7 - 5 = 2$.
(iii) these numbers $\quad 7 \times 2 + \ 1 = 15$
$$15 \times 1 + \ 7 = 22$$
$$22 \times 1 + 15 = 37$$
$$37 \times 1 + 22 = 59$$
are followed by dots (as markers) in the table.

The solutions of the original equation are therefore:

$$x = 1, 13, 25, 37, 49 \ldots \text{continue to add } 12$$
$$y = 2, 21, 40, 59, 78 \ldots \text{continue to add } 19$$

The *kuttaka* is an extremely sophisticated procedure and points to a deep understanding of number theory. Aryabhata did work of equal insight in other areas. He calculated the value of pi as $\frac{62,832}{20,000}$, that is, about 3.1416. (He also recognised that this was an approximation. It is, none the less, an improvement on Archimedes' value of pi as being 'between $\frac{22}{7}$ and $\frac{223}{71}$', that is, between about 3.1408 and 3.1428.) He laid a basis for trigonometry by developing a table of sines, which replaced Ptolemy's table of chords in astronomical calculations. Many centuries before Copernicus, he asserted that the rotation of the heavens was an illusion caused by the rotation of the

Earth about its axis: in other words, it was the observer who was moving, not the heavens.

Bhramagupta (598–c. 665)

Bhramagupta was born in Sind, now in Pakistan. He was a leading astronomer. His best-known works are the *Bhrama Sphuta Siddhanta* (a book on astronomy, which includes half a dozen chapters on mathematics) and the *Khandakhadya*, written in 665. His writings offer a definitive statement of the complexity and overriding excellence of Indian mathematics in this early period.

Bhramagupta provided the first systematic treatment of negative numbers and of zero, including definitive rules for multiplying positive and negative numbers, and for multiplying and dividing by zero. He accepted zero not only as a convenience in indicating an empty place in the number, but as a number in its own right. He provided a general solution for the quadratic equation. He recognised that it had two roots, and even that one of the roots could be negative. (For example, the equation $x^2 - 4 = 0$ has solutions $x = +2$ and $x = -2$.)

The recognition that a quadratic has two solutions, or roots, seems to have been passed over by the Arabs in their translations and development of Hindu algebra. It had to be rediscovered, over 1000 years after Bhramagupta's death, by Western mathematicians. Bhramagupta also gave the general solution for the linear (first-degree) indeterminate equation

$$ax + by = c$$

as $x = p + mb$ and $y = q - ma$, where p and q are any two solutions; different values of m (i.e. $m1, m2, m3$ and so on) give a set of values which x and y can take. He also proposed the indeterminate equation

$$x^2 = 1 - py^2$$

for solution. (This equation was later, erroneously, labelled 'Pell's equation' by Euler.)

Bhramagupta adopted the method of giving a number of solutions for indeterminate equations. Many of his examples are essentially the same as those of Diophantus (see page 87). This suggests that their work had a common source, such as Babylonian algebra. There is no evidence that the Greeks had any influence on Indian mathematics.

Pythagoras, as depicted by the French artist Thevet in 1584. The portrait shows that even 2,000 years after his death Pythagoras was still being (mistakenly) idealised in some quarters as a kind of scientific 'sage'. (*Mary Evans Picture Library*)

This engraving of the murder of Hypatia was made in early nineteenth century France. This highly romanticised interpretation shows Hypatia falling victim to brutish thugs. (*Mary Evans Picture Library*)

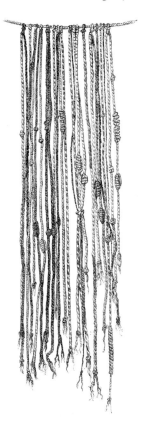

This *quipu*, now in the Berlin Folk Museum, dates from the height of Inca culture in Peru. It is made of waxed, plaited cords in several colours, and each colour, each cord, and each kind of knot has its own individual meaning (now, unfortunately, lost forever, without the *camoyoc* to explain it). (*Mary Evans Picture Library*)

The guests at this sixteenth century gaming party are rolling dice, betting on the result not with coins but with tokens. Such tokens are still used (for security?) in casinos and other gambling establishments throughout the world. (*Mary Evans Picture Library*)

Although the features in this portrait of Isaac Newton are claimed to be a true likeness, the artist has also idealised him as a scientific 'star'. (*Mary Evans Picture Library*)

This portrait of George Boole is thought to be a true likeness – and unlike the portraits of Pythagoras and Newton earlier, it certainly seems to show a real human being, rather than some personification of scientific 'genius'. (*Mary Evans Picture Library*)

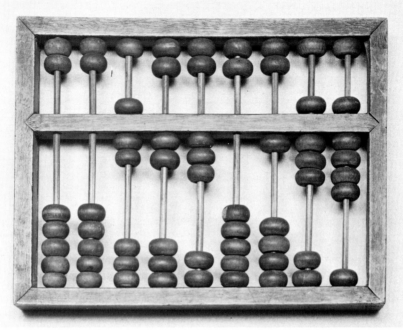

On this Chinese abacus, of unknown date, the upper rank of counters represents fives and the lower rank represents units. The figures shown, from left to right, are 0,0,7,2,3,0,1,8,9. (*E.T. Archive*)

This drawing, made in San Francisco in the 1880s, shows a Chinese merchant using an abacus for calculation. (The book on his right probably contains accounts.) The onlookers' expressions suggest that his actions seem magical, rather than simple and scientific – a standard reaction to such work throughout the history of calculation aids, from counting boards to computers. (*Mary Evans Picture Library*)

Photogravure portrait of Charles Babbage. The process (engraving based on a photograph) suggests that this, like the Boole portrait earlier, is an authentic likeness. (*Mary Evans Picture Library*)

The 'mill' from Babbage's analytical engine, as constructed under the direction of his son, Henry P. Babbage, and now in the Science Museum, London. In use, the machine would have had a wooden case, pierced with slots to show the figures on the dials. (*Trustees of the Science Museum, London*)

Bhaskara (1114–85)

Bhaskara was born in Bijjadha Bida in Mysore. Nothing else is known of him, except that in 1150 he wrote the *Siddhanta Siromani* ('The Gem of Mathematics'). This book is in four parts: *Leelavahti* ('the beautiful one'), on arithmetic, *Bijahanita* (algebra), *Goladhaya* (on the celestial globe) and *Grahaganita* (on the planets). The *Leelavahti* is written in verse with a prose commentary. It is an original work, mainly arithmetic, but with some geometry. There is a chapter on the *kuttaka* method (see page 119). Bhaskara is supposed to have written this text to distract his daughter from an ill-judged romance. The book bears her name, Leelavahti, and is written in an affectionate, charming style. A typical problem reads:

O tender girl, out of the swans in a certain lake, ten times the square root of their number flew away to Manasa Sarovar when the rains came at the monsoon. One-eighth went away to the forest called Sthala Padmini. Three pairs of swans remained in the lake engaged in amorous play. How many swans were there in all? [Answer: 144]

This way of presenting problems serves two purposes. It is imaginative and thus retains the interest of the student, without distracting attention from the essential facts of the problem. It demonstrates that mathematical methods are relevant to the solution of real-life problems (even when these are quite trivial). The student is trained from the early stages to select relevant data and ignore the persiflage. It is a humane approach which fastens on the student's other interests without demeaning his or her abilities or immaturity.

Another problem, from Bhaskara's *Bijahanita*, shows a similar, if less ornately-expressed approach:

In the interior of the forest, a group of apes equal to the square of one-eighth of their number play with each other. The remaining 12 apes are playing on a nearby hill. The echo of their play from the surrounding hills upsets the apes in the forest. How many apes were there? [Answers: 16 or 43 – both solutions are admissible.]

In a third problem, from the same book, the equation has two roots, but one is inadmissible:

A party of monkeys splits into two. The square of one-fifth their number less three went into a cave in the forest while the one remaining monkey climbed into a tree. How many were in the original group? [Answer: $x = 50$. The other solution ($x = 5$) is not viable.]

In the same book Bhaskara solves the cubic equation

$$x^3 - 6x^2 = -12x + 35$$

by rewriting it as follows:

$$x^3 - 6x^2 + 12x - 8 = 27$$
Therefore $(x-2)^3 = 3^3$
Therefore $x - 2 = 3$
Therefore $x = 5$

The Indian contribution

Our main problems in evaluating the Indian contribution to number arise from the lack of a continuing tradition of interest in, and study of, its development and history by Western scholars. The history of India has certain analogies with that of central America. The ancient civilisations of the Incas, Aztecs and Maya had arrived at about the same level as that of the Indians in wealth, material and spiritual culture. The Maya had also reached approximately the same stage in their work with number. (It is quite unclear where the original stimulus for this activity came from, whether from the Chinese or the Babylonians or from some other ancient culture.) But because the Christian religion effectively monopolised schools and universities in mediaeval Europe, Greek philosophy and mathematics became a dominant influence there. This meant that Western scholarship was defined for centuries in terms of knowledge of and concern for Roman and Greek learning. In the sciences, and especially chemistry, knowledge of Arabic, or of Arab books in translation, helped a few chosen individuals to counteract the provincialism and ethnocentrism of most of those who passed as 'learned'.

In much the same way as the Spaniards in the Americas decimated the population and unthinkingly destroyed the high culture of the Aztecs, Inca and Maya, and white settlers invading North America ravaged the indigenous Amerindian cultures, so the British, from the late 17th century onwards, annexed the subcontinent of India and took over all available resources. The East India Company and its servants proceeded to loot the Indian empire by private 'enterprise'. They had no interest whatever in understanding the achievements of Indian intellectuals, the detail of Indian religions, or any other specialised knowledge. They made no attempt to preserve and study the stock of intellectual materials in Indian universities. Disconcerted by the sensuousness of Indian art, and by Indian religions, because of their 'naturalism', they proclaimed that such things were contaminated by paganism. They refused to recognise or concede that Westerners might have something to learn from Indian scholarship, past or present.

THE MAYA

Nothing but flowers and songs of sorrow
Are left where once we saw warriors and wise men . . .
Have you grown weary of your servants,

O Giver of Life?

Aztec poet

The concept of civilisation

In the remote geological past, the surface of the Earth split into several land-masses. These drifted apart, as though floating in a liquid, to form the continents. Many millions of years later, in the largest of these land-masses, early human beings appeared and proceeded to develop various kinds of societies. By accumulated changes, some good, some bad, human society developed to the stage that historians define as civilisation. This was based on five crucial discoveries: how to control fire, how to plant seed and grow crops, how to tame and use work-animals like the dog, ox and horse, how to smelt iron and other ores to make tools and weapons, and how to use the wheel to move heavy loads. These discoveries, made over millennia in Europe and elsewhere, became so much part of everyday life as to be taken for granted. In the process, the arts and sciences were perfected. Human societies also kept changing their religions and philosophies. They invented nation-states and international wars.

Beginning in the 15th century, the 'benefits' of this 'civilisation' were imposed on the New World by explorers, soldiers and Christian missionaries. In America, they met many ancient civilisations, but in particular those of the Inca, Aztec and Maya. The model for these civilisations was totally different from others discussed in this book. They belonged to an entirely indigenous tradition, the Olmec–Maya–Toltec–Mixtec–Aztec form of ancient civilisation, which had grown up over millennia with reference not to the outside world but only to itself. It was in no way inferior to the culture of the European incomers – and only slightly more bloodthirsty. But the newcomers failed utterly to understand it, condemned its practices as satanic perversions and proceeded to plunder and destroy. Patterns of life and thought that had evolved over thousands of years were brought to an end in less than a generation by European (mostly Spanish) firearms and duplicity.

The main features of each of these cultures are known, but we can picture them best in terms of their high point, Mayan civilisation. There is a common pattern because each Indian civilisation, by right of conquest, annexed the achievements and culture of the earlier ones. Although the languages and the names of the gods differ, there is otherwise a close relation between the different culture patterns.

In contrast to the Old World, the New World nations had only two of the marks that European historians look for in defining a 'civilisation'. These were the control of fire and an economy based on cultivated crops. There were no horses, oxen or other domestic animals strong enough to serve as work-beasts. There were no metal tools or weapons (such as ploughs or metal swords), and no wheels except on a few toys (such as those recently unearthed in Mexico). Amazingly, the Central American Indians built cities, several times the size of the London of Henry VIII. They differed from European cities, being designed primarily not as centres of population but as vast religious complexes. Hereditary nobles and priests lived in these cities; artisans and tradespeople lived in the suburbs. The bulk of the population farmed the land and supported the ruling class by their labour and taxes.

The Maya had a written language (Mayan hieroglyphics), in which they composed books on their history, astronomy and astrology, religious rites and festivals, and on their culture in general. (This writing, including numbers, was also carved on memorial stones honouring rulers, their spouses and their children.) They produced artworks and jewellery which still dazzle connoisseurs. Their religion was no less adequate for the spiritual needs of the society, no less transcendentally satisfying, and far less intolerant, than the Christianity of the *conquistadores* and inquisitors who plundered them. Their work on the calendar and science surpasses that of European scholars of the same period. As far as arithmetic goes, they were about a thousand years in advance of Europe. Their observations in astronomy were not inferior to any world standards of the time.

Mayan hieroglyphics

Apart from the three manuscripts now in Madrid, Dresden and Paris, all written records of the Maya people were torched by the Franciscan monk

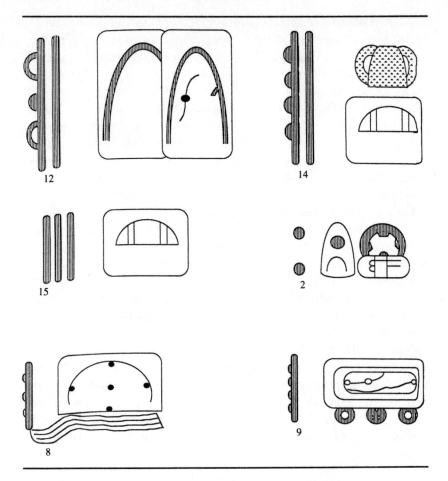

Diego de Landa (1524–79) one day in 1541 in Mani in Yucatan. Later, as Bishop of Merida, he did penance for his misdeed and spent the rest of his life collecting, by word-of-mouth from converted Mayans, all they knew about the culture he had destroyed, and writing it up in Spanish. Our knowledge depends partly on these records and on one or two accounts in Spanish written by later Maya converts. But mostly we rely on Mayan monuments: buildings and cities in Yucatan and Guatemala rescued by archaeologists from the jungle.

However, as one of these scholars remarked in passing, it is as if we were asked to describe North American culture with no sources other than half a dozen buildings, three prayer books and a copy of *The Pilgrim's Progress*. No account of Mayan language, hieroglyphics, astronomy or arithmetic is available except those composed after the Spanish conquest. The Mayan language is still spoken, in a modern form, by about two million people. In the last decade, after years of work, scholars have succeeded in keying this to the hieroglyphics. The result is a breakthrough in understanding similar to that which happened with Egyptian and Sumerian culture 170 years ago, when their writings were deciphered. Some 500 Mayan hieroglyphics, out of approximately 800, have been translated. Luckily for our purpose, many of these refer to the calendar, to astronomy and to numbers (which were de Landa's special interests, and therefore form the bulk of his reports). Mayan ways of thinking can also be recognised in Aztec writings, these being a little more accessible, and derivative from this source.

The unique Mayan script is known from inscribed stones at more than 120 sites in Central America. These include temple buildings, ceremonial centres, priestly residences as well as markers in the form of monumental stone pillars. These structures, intended to be commemorative, were invariably inscribed and dated. Dates were written in a standard form, in hieroglyphs. These can be extremely complicated, being made up of several different signs enclosed in a cartouche or boundary line. Many depict the heads of what archaeologists at first took to be gods, but are now known to include rulers, their wives and their children.

The realisation that the 'portraits' were of living people, not gods, was a recent and important breakthrough in working out what the hieroglyphics mean. Another major clue has come from computer analysis of the hieroglyphics. This has shown that many are dual in function, representing phonograms or speech sounds as well as ideograms or ideas. (Some 60 syllables have been identified as phonetic, giving a syllabary of the language.)

Mayan number and number symbolism

Any understanding of Mayan science and number is hampered by the absence of written texts to provide a background to their thoughts. We have bald statements of results; from these we have to deduce, in each case, the original problem, methods of working, algorithms and other facts. There is no record of the unknown geniuses who perfected the system nor any statement of how they did it. The account that follows, therefore, is more a summary of probabilities, based on what has been deduced so far, than an exhaustive catalogue.

In calculation, Mayans used the vigesimal system: that is, numbers were on a base of 20 (not 10 as in our decimal system). In this system, place-values go up in powers of 20. The first five place-values in the decimal system are 1, 10, 100, 1000, 10,000. The first five in the vigesimal system are 1, 20, 400, 8000, 160,000.

For text, the convention used to translate Mayan numerals into our Arabic figures is to write each number horizontally, with each place separated from the next by a dot. A vigesimal number, written in Arabic figures, would be expressed in the form: 1.10.2.14.3. Mayan hieroglyphic numbers, however, were written vertically, with the lowest denominations at the bottom, increasing as we move to the top. To discover what the number 1.10.2.14.3. stands for, therefore, we need to set out the numerals vertically (shown with their place values), as follows:

Mayan numerals	Place-value	Decimal value
•	1 times 160 000	= 160 000
≡	10 times 8 000	= 80 000
• •	2 times 400	= 800
•••• ≡	14 times 20	= 280
•••	3 times 1	= 3

Mayan numbers are read from bottom to top, and place-values or levels go up in 20s, not in 10s. The number 1.10.14.3, therefore, set out in Mayan notation in the left-hand column, is the total of all the numbers in the right-hand column, namely 241,083.

It is commonly believed that the Maya were obsessed by time. This may have been so. But it may equally be an exaggeration, a scholarly stereotype based on the fact that almost all we know about Mayan intellectual life revolves round their calendar, astronomy and astrology. (Once again, where would we start assessing a modern culture, if we had no information about it other than this?) Lacking proper historical records, students of Mayan culture necessarily become caught up in the study of the calendar and the numbers depicted on the monuments.

Mayan priests and rulers are thought to have worshipped time as a god. They pictured time passing in a stream. But this was not just a stream of abstract numbers (a common metaphor in many languages); in the Mayan image of time, each number was carried by a god. Each god had a distinct personality as well as a name. His or her temper fluctuated, just like a human's, between good and bad. These qualities affected the number associated with the god. It was also under the influence of two other gods. One was the patron of the day-sign, the other of the month to which the number was attached. This complex of influences gave an individual quality to each day. It created a continuing need for priests to read the past and predict the future, to declare the day-to-day emotional condition of the gods and to cast individual horoscopes.

Such ideas are difficult to relate to. Perhaps it helps to think of the 13 numbers of the sacred-month calendar as a moving sequence carried in file by gods who act as porters. Each porter deposits his or her number load at a kind of intermediate station, the day. When all loads are deposited, the file of gods moves on, returning to the starting point and picking up a new sequence of numbers. In this way, time rolls on and on, so to speak, carried by the gods in an unending procession.

This model also serves to explain the visible motions of the Sun and Moon. They too were guided in their courses, minute by minute, by divers divinities. In the darkness of night, the Sun and Moon still journeyed on in the Underworld, invisible in the darkness. There they were beset by evilly-disposed gods who sought to block their progress.

The heavenly bodies therefore needed human help. This was provided by sacred rituals. It might take the forms of self-mutilation, torture of others, even death. These were dues human beings must pay for the continuing survival of life and of the universe. Sacrifice and death in this world-drama were not demeaning to the victims. On the contrary, death in this cause was a privilege: it ensured immortality to those who were offered, or who offered themselves, as victims.

In most cultures, the passage of time is measured by studying the movements of the heavenly bodies, chiefly the Sun and Moon. These seem to circle the Earth in regular cycles; they have a clear relation to the eternal cycle of life, death and regeneration. This complex of ideas was taken over, or perhaps reinvented, by each New World nation in turn. For example, the Aztecs worshipped the Sun. They made themselves responsible for keeping the Sun god moving across the sky by feeding him the hearts and blood of human victims. The Mayan cosmogony was similar, if not so bloody. It involved observation of the Sun, Moon and several other planets, especially Venus, which they recognised as both morning and evening star.

To monitor the cosmic process, a number system is essential. This means different signs for different numbers. In more developed number systems, the numerals are in a definite order, each place standing for a power of the base. The base is a matter of use and convention, and may never even be mentioned. A sign for zero is needed to fill up any vacant places; this avoids confusing one place with another.

The first evidence for the Mayan number system dates from the 4th century, some 400 years before the codification of Hindu numerals and

Mayan numerals from 1 to 20

Name	Number	Sign
huu	1	•
ca	2	• •
ox	3	• • •
can	4	• • • •
ho	5	___
uac	6	•
uuc	7	• •
uaxac	8	• • •
bolon	9	• • • •
lahun	10	═══
tuluc	11	•
la-ca	12	• •
ox-lahun	13	• • •
can-lahun	14	• • • •
ho-lahun	15	≡≡≡
uac-lahun	16	•
uuc-lahun	17	• •
uaxac-lahun	18	• • •
bolon-lahun	19	• • • •
hun-cal	20	≣≣≣

more than a millennium before the zero arrived in Europe. The Mayan zero sign had two forms: a clam shell or a head, either full-face or in profile. (Although today we take the existence of a zero sign for granted, its invention was a most remarkable feat, which happened only two or three times in the entire history of humanity.)

Some numbers were, so to speak, more sacred than others. This was because they were 'nodes' of one kind or another, playing some special role. (A node is a special marker in a system, different in function and indicating some intended break-point from other members of the system.) For example, the fact that the base number was 20 made this number special, despite its mundane origin as the total of our fingers and toes. Another special number was five, the number of digits on each hand or foot. It was created by Hunab Ku, the senior divinity who made all things.

The number 13 was sacred. It was the original total of gods, and the basis of the sacred calendar. For the Maya, the world and sky were set out in 13 layers. Another number, 52, was sacred as the number of years in a bundle, rather like our century. The change-over between one group of 52 years and the next was a time when the very survival of the Earth and every living thing was pondered and decided by the gods. Yet another number, 400, was sacred as the number of gods of the night. There were 1600 stars visible to the Maya priests in the night sky. Each was a minor deity, conquered daily by the Sun god when he rose in all his glory. As Epictetus once remarked, 'Everything is full of gods!'

The Mayan calendar

The Maya recorded the passage of time in two ways: one was concerned especially with religious matters, the other with the daily round. The calender was a unique amalgam of the two.

First, was the sacred 'count-of-day'. For the Maya, the birthday or day-sign determined each person's fate through life. It connected the new-born with the god of the day. The child remained under the influence of this patron all his or her life. Each god had two aspects, one malevolent, the other benevolent. On balance, some had goodwill to humans, others the reverse. In the first case, the child was lucky to be born under the patronage of a god who wished him or her well. In the second case, he or she had to ensure that the bad-luck god was deferred to, and propitiated throughout life, especially during vulnerable periods. The last five days of each year were especially dangerous. No work was done and everyone stayed indoors. The priests could interpret the sacred calendar and its meaning for each person every day. It was therefore wise to be guided by their counsel.

The calendar had a social reference as well as an individual one. Religious rites and festivals were linked to seasonal events and were under the domination of the Sun and Moon. The calendar took notice of both sacred and secular events to specify the date. There was a four-fold index defining each day in a long sequence of time: the number of the day in the sacred calendar; the god of its day-sign; its number in the year-calendar; and finally, the god of the month in which it fell. (Compare the four pieces of information in a modern dating system: for example 'Thursday, 20 March 1990' means 'the god Thor's day, the 20th of Mars' month, 1990'.)

In the Mayan ritual calendar, the first number identifies whether the day was one of the numbers between 1 and 13. Second was the name of the god, one of 20 day-signs. Since each of the 13 days was connected in turn with one of the day-signs, it was possible to identify 260 days (13 × 20) by a number and a sign. So the day's position was identified by its place in the *tzolkin* or sacred year, as in the chart below.

Day-name	Day-number												
ik	1	8	2	9	3	10	4	11	5	12	6	13	7
akbal	2	9	3	10	4	11	5	12	6	13	7	1	8
kan	3	10	4	11	5	12	6	13	7	1	8	2	9
chicchan	4	11	5	12	6	13	7	1	8	2	9	3	10
cimi	5	12	6	13	7	1	8	2	9	3	10	4	11
manik	6	13	7	1	8	2	9	3	10	4	11	5	12
lamat	7	1	8	2	9	3	10	4	11	5	12	6	13
muluc	8	2	9	3	10	4	11	5	12	6	13	7	1
oc	9	3	10	4	11	5	12	6	13	7	1	8	2
chuen	10	4	11	5	12	6	13	7	1	8	2	9	3
eb	11	5	12	6	13	7	1	8	2	9	3	10	4
ben	12	6	13	7	1	8	2	9	3	10	4	11	5
ix	13	7	1	8	2	9	3	10	4	11	5	12	6
men	1	8	2	9	3	10	4	11	5	12	6	13	7
cib	2	9	3	10	4	11	5	12	6	13	7	1	8
cahan	3	10	4	11	5	12	6	13	7	1	8	2	9
eznab	4	11	5	12	6	13	7	1	8	2	9	3	10
cauac	5	12	6	13	7	1	8	2	9	3	10	4	11
cahau	6	13	7	1	8	2	9	3	10	4	11	5	12
imix	7	1	8	2	9	3	10	4	11	5	12	6	13

This 'sacred year' calendar was a precise instrument for arranging state affairs in accord with the wishes of the gods, as declared by the priests. It is still not known what the 260 days refer to. There is no clear relation between this number and anything happening in the heavens, the normal source of calendar cycles.

The sacred calendar is, however, only one part of the story: half a calendar, so to speak. The other half, the secular calendar, was known as *haab* by the Mayan priests and as the 'vague year' by Western scholars. This had to do with the seasons and agriculture. It was based on the solar cycle, the time it takes for the Earth to go round the Sun. This is now known to be 365.2422 days, but the Mayans only used 360 days – hence the European name 'vague year'.

In conformity with the Mayan vigesimal counting system, each vague-year month was reckoned to contain 20 days. This meant that there were 18 months of 20 days each, giving a total of 360 days. The five days over were grouped as a final 'month' of five days of special danger and bad luck. This left one-quarter of a day (actually 2422 ten-thousandths of a day) over. The Gregorian calendar solves this problem by a complicated device of

Tzolkin Calendar		Haab Calendar	
day-numbers (days 1–13)	day-names (20)	day numbers (0–19 days)	month names (18 + 1)
1	imix	0	pop
2	ik	1	uo
3	akbal	2	zip
4	kan	3	zotz
5	chicchan	4	tzec
6	cimi	5	xul
7	manik	6	yaxkin
8	lamat	7	mol
9	muluc	8	chen
10	oc	9	yax
11	chuen	10	zac
12	eb	11	ceh
13	ben	12	mac
1	ix	13	kankin
2	men	14	muan
3	cib	15	pax
4	cahan	16	kayab
5	eznab	17	cumhu
6	cauac	18	uayeb
7	cahau	19	—

declaring certain extra days to be leap-year days which are added or not according to a very complex pattern. The Maya used a similar solution, and also made adjustments (no less complex), based either on observation of Venus or on solar eclipses.

As we have indicated, to identify a day in the Mayan calendar, it was necessary to state four components:

number, tzolkin month; number, haab month.

It is as if each of these components is displayed on a separate dial, behind which is a continuous paper strip. As the days pass, each strip is moved on one position. The *tzolkin* (sacred) calendar works by moving a day at a time in the pattern shown in the previous table. The *haab* (everyday) calendar also moves a day at a time. But it runs from zero (which is the first day) to 19 for each of the 18 regular months of 20 days, and then for only five days of the 19th, 'fateful' month.

The 'long count'

Mayan priests made decisions about the dates of sacred events and for the agricultural round. There was no pressure for a system that could be understood by ordinary people, quite the reverse. This explains the complexity of the calendar. They did not use either the 365-day count or the 260-day count alone for dating events. These numbers and names were combined to point the particular day and month. However, the dates inscribed on monuments for state events, such as the death of a monarch, gave the year as well. This was done by means of the 'long count'.

The complexity of the system stems in part from the Mayan creation-myth. In this myth, the gods were dissatisfied with their first four attempts to create human beings. Each time, they destroyed the whole species and began again. They used different materials each time, experimenting, for example, with mud and monkeys. The fifth material was dough, made from maize flour and water. This 'took', but we are still on probation; the gods may destroy the world again. The big decision will be made on a date which is, in the modern calendar, 24 December 2011 (but don't mention it!).

This chequered history of creation means that the calendar begins not at day one, but at a date written as 7.0.0.0.0., that of the gods' fifth attempt at creation. (It corresponds in our calendar to 3133 BC.)

A second complication in dealing with the calendar is that the vigesimal system is slightly altered in 'sacred calendar' dates. (Sacred arithmetic is slightly different from secular.) In the vigesimal system, as has been said, numerals refer to the different powers of 20. The power is decided by the position of the numeral in the number. The basic unit of the calendar is, of course, one day. In Mayan, a day is called *kin*. The next place-value (moving to the left) is called a *uinal*. However, for sacred-calendar purposes, the *uinal* is reckoned not at 20 *kins* but at 18. This change is made to give the 'vague year', which contains only 360 days. The other places then move up by 20s, as in the normal vigesimal system. Thus, periods of time are counted as:

> 20 *kins*   = 1 *uinal*   (this is the normal value)
> 18 *uinals* = 1 *tun*   (360, not 400 days)
> 20 *tuns*   = 1 *katun*   (7,200 days)
> 20 *katuns* = 1 *baktun* (144,000 days)
> 20 *batuns* = 1 *pictun* (2,880,000 days)

The system continues by 20s, through *pictuns* to *calabtuns*, *kinchiltuns* and *analtuns*. Thus the time scale of the 'long count' can span a period of 367,000,000 million years or so.

In terms of the Western calendar, Columbus reached the New World on 13 October 1492. The table below shows how this date would have been worked out, and recorded, in Mayan terms:

Gregorian date	Event	Mayan date
3133BC	Creation (Mayan)	7.00.00.00.00
AD 1492	Columbus in America	4.14.12.06.04
4625 years = 1,689,244.25 days		11.14.12.6.4 = 11 baktuns, 14 katuns, 12 tuns, 6 uinals, 4 kins

Counting 286 days to 12 October, the Mayan date is

            11.14.12.6.4   4 cimi 14 pop

If we take the calendar as an index of the mathematical achievement of a pre-modern society, there is an almost exact parallel here between the Maya and mediaeval Europe. In both societies, calculations were the preserve of priests, and the calendar chiefly existed to set exact dates for rituals or ceremonial events. (The work of Bede (9th century) in this regard, stabilising and codifying the Christian Church calendar, mirrors that of his Mayan counterparts.) Because no instruments had yet been invented, both groups operated on the basis of naked-eye observation of the heavens. Both groups were hampered by the need to harmonise human numbers with the statements of religious myth. Both groups solved problems by making adjustments which subsequently became inextricably part of the canon. Such practices made little difference, so long as knowledge of the calendar remained a sacred mystery, tended and interpreted, by a specialist élite.

THE ARABS: RENAISSANCE OF NUMBER AND SCIENCE

Baghdad and Cordoba, the Eastern and Western Arab Caliphates . . . [were]
like the two terminal points of a gigantic intercontinental system . . . between
which the intellectual current . . . flowed . . . through the superconductive
cable of a single Arabic language . . . The flow was from East to West because
– to carry on the metaphor – in general the Orient was the transmitter and the
West the receiver.   *Karl Menninger*

AT the time of Mohammed's birth (AD 570), the Arabs were a nomadic,
pastoral, desert people, speaking a Semitic language, and at a similar
cultural level to their ethnic relatives the Jews. A century later, they had
begun a huge campaign of conquest which led to cultural dominance not
only in the countries bordering the Red Sea, but throughout the Middle
East and in large parts of Africa, Central Asia and south-western Europe.

Although early Muslims, like Jews and Christians, were intolerant of
other religious revelations, in other respects the Arabs were assimilative
and intellectually adaptive. They were eager to absorb, improve and
transmit the culture and science of all the countries that fell under their
sway, and to learn all they could from such ancient civilisations as those of
Egypt, Babylon and India. Their scholars studied Western science,
translating the Greek and Latin texts. Their declared object was a rebirth of
human knowledge. But they also made their own contribution, developing
new subjects of study such as chemistry, algebra and trigonometry.

The golden age of Muslim science and art coincided with the European
Dark Ages. With the fall of the Roman Empire in the 4th to 5th centuries,
apart from a few metropolitan centres, Europe settled back into the tribal
barbarisms which, it might have been imagined, 1200 years of Roman rule
would have 'civilised' out of existence. Led by their churchmen, the
Europeans were trapped in a primitive fundamentalism, with its resulting
intolerance and persecution of secular knowledge. The authority of the
Bible, and of a few favoured 'pagan' scholars such as Aristotle, strangled

not only any new knowledge that seemed to contradict revealed truth, but the spirit of inquiry itself.

Over some 700 years, therefore, from about the 7th to the 15th century, Arab architecture, art and literature outshone those not only of the Christian nations of Europe but of their Greek and Roman sources as well. Although the Arabs were ferocious in war, in peace their life-style was tolerant and civilised. They brought to the arts, and to science in particular, a no-nonsense practicality which rescued intellectual activity from Greek frivolity and Roman intolerance and which continues to permeate scientific study to this day.

Throughout their vast dominions, the Arabs set up libraries, observatories and research institutes. Taking as their authority the phrase in the Koran, 'He hath created the heavens and Earth to set forth his truth' (Sura xvi, 3), Arab scholars aspired to record every piece of knowledge gained by humankind and to advance it further. They developed vast programmes for publishing their own scientific and mathematical works and for translating treatises from Syrian, Persian, Chinese, Greek and other languages. More than any other investigators, in any culture before them, they exploited the test of practice and experiment in the quest for scientific truth. They may have lacked the hyperactive imagination of the Greeks, but they amply compensated for it by thoroughness and pragmatism.

This genial view of the Arabs was not shared by most Christians of the time (whose attitude still colours our Western outlook on Islam in particular and Arab culture and character in general). From the time of the Crusades on, Christian rulers (including the Pope), who were motivated as much by mercantile covetousness and rivalry as by their proclaimed religious mission, wrote off the Arabs as minions of Satan, to be shunned, or if possible eliminated, wherever they were found. The Arabs' interest in science and scholarship did nothing to soften their unacceptable image in a Europe gripped equally by military adventurism, religious exclusivism and feudal law.

Al-Khwarizmi (c. 680–750)

Little is known about the life of Abu Abdullah Mohammed ibn Musa al-Khwarizmi al-Magusi, possibly the most influential of all Arab mathematicians. His name tells us that his family was from Khorezm in Persia (now in the USSR), and that at least one of his ancestors was a Magus, or priest of Zoroaster. Al-Khwarizmi worked in the House of Wisdom in Baghdad, a centre for scientific research and teaching set up by the Kalif al-Mamun (son of the Khalif Harun al-Rashid, who was immortalised in the *Arabian Nights*).

Al-Khwarizmi's works had a profound bearing on the development of world mathematics. The translation of his book on arithmetic, for example, introduced Arabic numbers to the West, setting in train a process that led to the use of the nine Arabic numerals, together with a zero sign, as the most basic tools of science. His book on algebra both gave a name to this branch of mathematics and carried the subject forward far beyond its primitive beginnings in Diophantus, the 4th-century Alexandrian (see page 87). Al-Khwarizmi's stated intention in writing the book was to enable scholars to solve complex practical problems, for example, calculating the division of an estate between the legitimate heirs in accordance with Islamic law (which bound the testator to make dispositions for his spouses, sons and daughters, brothers, nephews and nieces, in stated portions according to their degree of relationship to the donor). But al-Khwarizmi went further than such mundane matters. He was also interested in the theoretical aspects of algebra as the science of equations.

Al-Khwarizmi's arithmetic

Al-Khwarizmi's treatise on arithmetic was the first book ever to explain the operations of the decimal numerals. Its original Arabic text has been lost, and the treatise survives chiefly in translation. In 1857, a copy of a 13th-century Latin translation was discovered in the University of Cambridge library. It begins with the words, 'Dixit Algorismi . . .' ('Thus said al-Khwarizmi . . .'). This opening signals the fact that al-Khwarizmi provides a new start in the study of mathematics. It also gives a new word ('algorism'), used first for the body of knowledge now known as 'arithmetic'. Al-Khwarizmi healed the rupture effected by the Greeks between number theory (which they called *arithmetika*) and practice (which they called *logistika*). In Christian Europe, arithmetic was known during the Arab supremacy as 'algorism', 'augrism', or by other variants of al-Khwarizmi's name. Many students of the subject believed that Algorism was a king or prince who decreed that calculations must be carried out according to the laws ('laws', not 'rules') set out in this text. In modern times, 'algorithm' has a more restricted meaning: a specific routine for solving a particular problem. For example, we learn algorithms for doing addition, for long division, or for finding the square root – and most of them are to be found in al-Khwarizmi's text.

The use of the nine Arabic numerals and the zero in al-Khwarizmi's work became the centre of a three-centuries-long ideological battle in Europe for and against the new arithmetic. Forces of change supported the Arab programme: the decimal place system, the use of 10 symbols only to represent all numbers. Opposing change was the majority of merchants and accountants, accustomed to the use of the abacus and of Greek and Roman alphabetical numerals. Counters, such as small discs about the size of coins, were well established for number work.

Arab numerals and algorithms made arithmetic so simple that one could throw away all auxiliary aids such as the abacus, and work directly with the numbers themselves. This made it easier to grasp their nature as abstract entities that might stand for any set of concrete objects, or that could be dealt with as pure abstractions. Euclid and other Greek mathematicians had freed geometry from the shackles of land surveys and building problems, enabling scholars to think about the abstract properties of space. The Arabs did a similar service for number.

Even with the help of paper, pen and ink (all scarce commodities in the pre-modern world), the unwieldy alphabetic numerals did not lend themselves easily to simple calculation. 'Sums' more complex than addition and subtraction were ruled out. As a result, attention was switched to the numerous and complex relations inherent in the natural-number system. For example, Pythagoras and his associates spent thousands of hours trying to figure out the connections between the first 10 numbers. The fact that $1 + 2 + 3 + 4 = 10$ (the *tetraktos*) gave them subject matter for endless days, weeks, even years of fascinated meditation (see page 82).

The Arabs knew the effects of such scientific shamanism from their own earlier encounters with numbers. Like other Semitic nations, such as the Hebrews of the Old Testament, they originally used letters of their alphabet as numerals. In earlier days they also used the sexagesimal system learned from ancient Sumeria and Babylon. Clearly, these experiences prepared them for discovering the immense potential of a fully-fledged decimal system. By using only 10 symbols, instead of the earlier 60, all calculation could be speeded up and simplified.

The invention of the zero sign in particular ('zero' from the Arabic *tzifer*, 'empty') transformed calculation from a concrete to an abstract art. It made place-value as crucial an indicator of the meaning of a numeral as its physical appearance. The true value of each numeral could be found only by combining its face value with its place-value, as shown in the extended number. The place-value was obtained by counting the number of the place, from right to left. The place number, starting the count on the right, represented units, followed by the tens' place, then the hundreds'. The power of 10 (hence the name decimal) was the value of that position in the number. Thus the same numeral, say 7, could stand for seven units, seven

Arabic alphabetic numerals (Gumal)		
(1–9)	٩ ٢ ; ، . د ٤ ب ١	(units)
(10–90)	ص ق ف ع س ن ل ك م	(tens)
(100–900)	ك ض ذ خ ث ت ش ر ق	(hundreds)
(1000)	غ	(1000)

(Read from right to left)
This system was replaced, over a long period, by the decimal numbers:

East Muslim: ١ ٢ ٣ ٤ ٥ ٦ ٧ ٨ ٩ .

West Muslim: 1   2   3   4   5   6   7   8   9   10

tens, seven hundreds, seven thousands, and so on, depending on exactly where it was placed in the number. When the value of each numeral was multiplied by its place-value, the sum of these results gave the value of the number. For example, 777 stands for 7 units plus 7 tens plus 7 hundreds or, from left to right, $700 + 70 + 7 = 777$.

We have learned these facts about the decimal system so completely that we accept them as commonplace, not worth mentioning. Indeed, we think of the 10 decimal numbers as 'natural' or even 'ordained'. For centuries, however, mediaeval Europeans found the notation so mysterious and difficult that they declared Arab mathematics Satanic in origin, and believed that those who practised the new skills were magicians and swindlers. (The history of the metric system in English-speaking countries provides an interesting parallel. Invented by the French in the 1790s, the metric system has still not been fully accepted, except in scientific work, and is regarded by many people with suspicion, even fear, as if the numbers themselves are threatening.)

There is no title page for al-Khiwarizmi's treatise on arithmetic, but lists of books in Arabic libraries refer to it as 'The book on Addition and Subtraction by Indian methods'. This gives the clue both to its contents and its method. It is a synthesis, explanation and refinement of existing knowledge, most of it Indian. After a detailed account of the decimal system and the use of the zero sign, and a section on how to communicate without error such large numbers (used in astronomy) as 1,189,703,051,492,863, al-Khwarizmi goes on to describe in minute detail what he calls 'Hindi' methods for the basic operations of arithmetic. Addition and subtraction are performed in the same way as we do them. Multiplication is done by the 'lattice' method (Arabic, *shabakah*). (This is much the same as Napier's bones, described on page 166. Indeed, Napier almost certainly borrowed the idea from the Arabs). It can be demonstrated in the following example, showing how to multiply 932 by 567:

Method   First draw a pattern of squares, three by three. (Three is the number of figures in 567 (the multiplier) and in 932 (the multiplicand), respectively). Then draw and extend diagonals to form a lattice, as shown. Multiply 5 by 9, writing the answer, 45, on the lattice on the first row, below the 5, with the tens figure above the diagonal. Then multiply 6 by 9, writing the answer in the same way below the 6. Do the same for the 7. Then multiply each of 5, 6 and 7 by 3, writing them in the middle row and making sure to separate the tens from the units as before. Lastly, multiply by 2, putting the results in the bottom row. Then add the figures diagonally, units to units and tens to tens, placing the totals in the diagonal lattice at the bottom. When the total of a diagonal is more than 9, some extra carrying is necessary. This is shown by the arrows in the diagram.

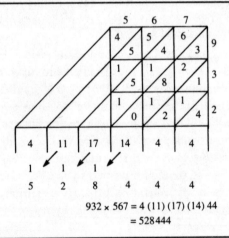

$$932 \times 567 = 4 \, (11) \, (17) \, (14) \, 44$$
$$= 528444$$

Division is equally simple. Let us, for example, divide 17,978 by 472. We set it out on a grid, as shown in the table below. Since there are five numerals in the dividend, we rule off five columns vertically. The dividend

1	7	9	7	8
0	5	9	7	8
0	3	8	7	8
0	3	8	1	8
0	0	6	1	8
0	0	0	5	8
0	0	0	4	2
		4	7	2
0	3	8		

$17\,978 \div 472 = 38$ remainder 42

(17,978) is written on the top line, the divisor (472) on the second line from the bottom; the result, or quotient, goes on the bottom line. Starting with 4 (the first figure of the divisor), we find that it does not divide into 1 (the first figure of the dividend), so we put a zero in the answer. Four divides into 17 four times, but on multiplying the divisor with this trial number, 4, we find the product to be too large. So we select 3 as our first figure after the zero in the quotient. The divisor is then multiplied by 3, the result being subtracted from 17,978 in three steps ($3 \times 4 = 12$; $3 \times 7 = 21$ and $3 \times 2 = 6$). Since we follow the frame in laying out these intermediate steps, we have no need to think about place values. The result of these three subtractions is to leave 3878. This is now divided by 472, giving 8 as the second figure in the quotient. As before, the subtraction is carried out in three steps ($8 \times 4 = 32$; $8 \times 7 = 56$ and $8 \times 2 = 16$). On subtracting, the final remainder is 42.

(We could, of course, perform the division more speedily by compressing the three steps into one. But this would increase the chances of error because 'carries' would then have to be made mentally. Doing it stepwise turns it into a purely mechanical routine: the algorithm for division.)

The Arabs are sometimes criticised for dealing only with 'easy' problems. The statement betrays an inability to understand the nature of their contribution. Running a mile in four minutes is now almost routine for Olympic athletes. But when first mooted as a project, it was said to be impossible. Similarly, breaking the sound barrier by flying a plane faster than 18 miles per second was believed to be so fraught with danger as to be impossible. As soon as the feats were achieved, they were repeated and repeated until they became almost humdrum. Exactly the same is true of al-Khwarizmi's algorithms. The fact that they now seem so easy is due to our familiarity with them. But that does not detract from their originality and importance at the time. We are where we are now, intellectually speaking, because they were there then.

Fractions

A whole chapter of al-Khwarizmi's book deals with fractions. The connection, in English, between the words 'fraction' and 'fracture' also exists in Arabic. *Kasra*, meaning 'broken numbers', is al-Khwarizmi's word. He describes how a single unit can be divided into pieces by each numeral of the decimal system, to give fractions which have special names, as follows:

$\frac{1}{2} = nisf$	$\frac{1}{7} = sub$
$\frac{1}{3} = tult$	$\frac{1}{8} = tumu$
$\frac{1}{4} = rub$	$\frac{1}{9} = tus$
$\frac{1}{5} = hums$	$\frac{1}{10} = ushr$
$\frac{1}{6} = suds$	

This introduces a discussion of sexagesimal fractions. They were written from left to right and described as 'seconds', 'minutes' and 'degrees'. We still use these terms in measuring time, arcs of a circle, or angles. The connection is with the apparent motions of the stars. As seen from Earth, the heavenly bodies appear to revolve with uniform speed in circular orbits, and in a clockwise direction. This uniform motion explains how the apparent motions of stellar groups or the star constellations became one of the earliest ways of measuring time. It also explains the design of clock-faces.

Sexagesimal fraction numbers are read from left to right. Thus, in the number 3. 24. 36. 48., the first figure, 3 stands for three units; 24 stands for $\frac{24}{60}$; 36 stands for $\frac{36}{60 \times 60}$ and 48 stands for $\frac{48}{60 \times 60 \times 60}$. The number would be read off as 3 units, 24 degrees, 36 minutes, 48 seconds. This is equal to 3.4102 in decimals (except that it is a repeated fraction, in which the final 2 is repeated ad infinitum. If we wanted a precise value, it would be written as $3 + \frac{3692}{9000}$).

Al-Khwarizmi also gives a method for multiplying unit fractions together. Suppose, for example, that the problem is to multiply $(8 \frac{1}{2} \frac{1}{4} \frac{1}{5})$ by $(3 \frac{1}{3} \frac{1}{9})$. Al-Khwarizmi would set this out in a table, as below.

We have to multiply:

$8\frac{1}{2}\frac{1}{4}\frac{1}{5}$ by $3\frac{1}{3}\frac{1}{9}$

The solution is written as: *Commentary:*

40	1 080	27
358	33 294	93
	30	
	894	
	1 080	

Common denominators
Converting numerators to
these three denominators

$8\frac{1}{2}\frac{1}{4}\frac{1}{5}$

(i) $\frac{320}{40}\ \frac{20}{40}\ \frac{10}{40}\ \frac{8}{40}=\frac{358}{40}$

$3\frac{1}{3}\frac{1}{9}$

(ii) $\frac{81}{27}\ \frac{9}{27}\ \frac{3}{27}=\frac{93}{27}$

Multiply (i) by (ii):

$\frac{358}{40}\ \frac{93}{27}=\frac{33\,294}{1080}$

$=30\frac{894}{1080}$

We still use al-Khwarizmi's algorithm, but in simpler (numerical) form:

$8\frac{19}{20}\times3\frac{4}{9}=\frac{179}{20}\times\frac{31}{9}=\frac{5549}{180}=30\frac{894}{1080}$

The method involves multiplying all the denominators together to find the common denominators (40 for the one, 27 for the other). We then express the whole numbers as fractions to these denominators. (We then separately add the two sets of fractions, having first expressed each set in terms of their common denominators (the two sums are $\frac{358}{40}$ and $\frac{93}{27}$). The two fractions are then multiplied together in this form, numerator by numerator, denominator by denominator ($358\times93=33,294$; $40\times27=1080$). The answer is thus $\frac{33,294}{1080}$. By division, we convert this to a mixed fraction: $30\frac{894}{1080}$.

In another section, al-Khwarizmi gives a method for multiplying mixed numbers (that is, whole numbers and fractions). This time, we are working with what we would call ordinary fractions: those whose denominators may be more than one. He gives as an example $3\frac{1}{2}$ multiplied by $8\frac{3}{4}$. This is done as follows:

3	8	whole-number parts
1	3	numerators
2	4	denominators
6	32	whole-number parts multiplied by denominators
7	35	above numbers plus numerators
$\frac{14}{4}$	$\frac{35}{4}$	more multiplying to give same numerators
Answer: $\frac{490}{16}$ that is, $30\frac{5}{8}$		product of above numbers

These answers have been modified only slightly from al-Khwarizmi's originals, by drawing lines and giving verbal explanations. It can easily be seen how difficult the new arithmetic must have seemed to anyone trained in older methods – and also, once you grasp the principles, how much quicker and more versatile it is.

Al-Khwarizmi's algebra

Al-Khwarizmi's book on algebra exists in the original Arabic, and in several Latin versions. The word 'algebra' actually comes from its title, *A Brief Account of the Methods of al-Jabr and of al-Muqabala*. *Al-jabr* (literally 'transformations') refers to how the balance of an equation is maintained when moving unwanted positive or negative quantities from one side to the other: by changing their signs and moving them to the other side. The word *al-muqabala* means the method of dividing every term of the quadratic equation by the coefficient of the second-degree term.

These two methods, *al-jabr* and *al-muqabala*, were the first steps of the algorithm that al-Khwarizmi developed for solving quadratic equations. An example will show how the routine works. Suppose that you cut, from a single piece of carpet of unknown width, and of length 10 units, a strip of area 21 square units, to leave a square, and you wish to know what the width of the carpet was. You begin by making a drawing of the carpet showing the various sizes. The unknown quantity is the width. Al-Khwarizmi calls this the 'root' (today, we use this word in a more restricted sense). The root is of the first degree. When we square it, that is multiply it by itself, the answer will be an area. This is the unknown ('root') raised to the second power and is called the 'square'.

We have to find an equation in one of al-Khwarizmi's five standard forms (see page 148). To make it simpler, we have substituted W (standing for 'width') for root, and $W.W$ ('W times W') for square.

Using symbols, not words, the equation can be written:

$$W^2 + 21 = 10W.$$

Al-Khwarizmi says, 'We now have a quadratic equation in one of the forms for which we have devised a routine. It is of type four, namely, squares (W times W) and numbers (21) equal to roots ($10W$). To solve it the routine is:

'Divide the number of roots (10) by two . . . . answer   5
Multiply 5 by itself. . . . . answer 25
Subtract the number, $25 - 21 = 4$ . . . . answer   4

Carpet	Equation
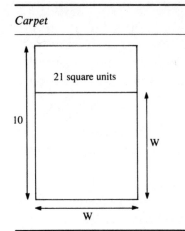	The problem is defined in terms of the equation: $W.W$ plus 21 equals $10W$

Take the square root of this last . . . . answer 2

Subtract from half the number of roots $(5-2)$ . . . . answer 3

3 is one of the roots you seek; the square is 9'.

A similar routine is given to find the second root:

'Then $9+5=14$; Divide by 2. . . . . answer 7
7 is the other root you seek; the square is 49.'

Thus the carpet could have had a width of either 3 units or 7 units. (It is impossible, without more information, to find out which.) It is part of the routine that, when al-Khwarizmi solves a quadratic, he reports the root and also the square. We now take it for granted that the 'root' is the required solution and that the 'square' is implied.

We see from this case that al-Khwarizmi is aware that a quadratic has two positive solutions (in this case $W=7$ and $W=3$). He knows that some equations have a negative solution, but he never quotes it. This is because, in his day, the negative solution had no meaning. It could never, for example, be the answer to a carpet problem: no carpet could ever be (say) -3 units wide.

Al-Khwarizmi's theory of quadratic equations

Al-Khwarizmi says that there are six kinds of equations. Five of these are quadratics, the other is the linear equation. (Linear means an equation of the first power of the unknown; such an equation can be represented by a straight-line graph, hence 'linear'.) He provides a method for solving each type. As we have seen, he speaks of the unknown quantity as the 'root'.

In quadratic equations the highest power is two, the root raised to the second power (*quadra* is Latin for 'a square'). Al-Khwarizmi calls it simply the 'square'. In solving these equations, his method is: first, divide by the number of squares. His algorithms depend, as we would say, on the coefficient of x^2 being 1. He realises that quadratics have two solutions, but only deals with the case where they are both positive. He ignores zero and negative solutions if they occur.

It was many centuries before negative numbers were understood simply as the opposites of positive numbers, with a meaning in terms of (say) direction, or as a debit balance in an account. Neither could al-Khwarizmi deal with the square root of negative numbers (these are now called 'imaginary numbers'). They were assigned a physical meaning only when alternating currents were discovered by Faraday in the 1830s. (Imaginaries have to do with the properties of alternating, as opposed to direct currents.) These various categories of number were perceived as numbers, only when they could be related to other parts of a number system. They then became amenable to the ordinary laws of arithmetic. Fractals (see page 241) are compounded of complex numbers, with a 'real' and 'imaginary' part.

Al-Khwarizmi's major contribution to this total picture was to the theory of equations. As we have said, he organised this area of mathematics by recognising six types of equation. These were as follows:

1. squares are equal to roots, as in the practical example: $x^2 = 8.x$
2. squares are equal to a number, as in the practical example: $x^2 = 4$
3. roots are equal to a number (linear equations): practical example: $8x = 4$
4. squares and roots are equal to a number, as in the practical example: $x^2 + 4.x = 12$
5. squares and numbers are equal to a number, as in the practical example: $x^2 + 4 = 5.x$
6. roots and numbers are equal to a square, as in the practical example: $5x + 6 = x^2$

(NB. The dots in these expressions are multiplication signs.)

Each of the five quadratics here (no. 3 is linear, not quadratic) has its own special method of solution. These all depend on *al-jabr* and *al-muqabala*, and are all solved by drills similar to those shown above.

Omar Khayyam

Gheyas ad-Din Abu al-Fath Omar Khayyami was born in 1048 in Nishapur (now in Iran), and died there in 1122. He spent much of his life as a wandering scholar. By the time he was 26, he had worked in observatories in Samarkand, Isfahan, Rey, Merv and other towns in central Asia. In middle life he was awarded the position of court astronomer to Sultan Alp Arslan, and was able to spend the rest of his life in his own home town, following the pursuits of astronomy, mathematics and poetry.

While in Samarkand, Omar wrote a book on algebra much like al-Khwarizmi's. Later, he wrote a commentary on Euclid and a treatise on methods of finding square roots and other roots of numbers. This last has been lost. Among Arab scholars, he is best known for his reform of the calendar, undertaken while he was in charge of the royal observatory at Isfahan. He drew up astronomical tables, and made use of his own and Babylonian observations. His work was so precise that it was only in error once every 5000 years.

The sultan had no interest in intellectual matters. He preferred the political arena and the military arts, leaving the cultivation of the other arts and sciences to his vizier and the team of scholars. Omar cast horoscopes for his royal master on request, but was sceptical about their ability to predict the future. He was not averse to predicting the weather for the sultan (who needed guidance about which days were good for hunting).

The only break in Omar's placid routine came when the sultan died in 1092, and the widowed sultana closed the observatory and stopped paying his salary. Her advisers had accused Omar of holding sceptical and rationalist views. He at once performed the pilgrimage to Mecca, indicating that he was a sound Muslim believer, and was reappointed. He remained in office until his death in 1131. He was an ornament to an otherwise undistinguished court, writing on philosophy (in the course of which he disagreed with Aristotle), Islamic law, history, medicine, astronomy and mathematics. Of these, only his work on algebra, a few pieces on philosophy, and his commentary on Euclid have survived. In English-speaking countries he is best known as a poet, at least since 1859, when his verses were translated by Edward Fitzgerald.

Omar's poetry, the *Rubai'yat*, was written as a respite from official duties and mathematical research. The *rubai* is a four-line verse. Its first two lines are like the major and minor premises of a syllogism in logic, stating indisputable truths. The third line states either a moral dilemma that contradicts the accepted position, or a conclusion from the premises. The fourth line repeats this conclusion with great emphasis. (Fitzgerald's English does not always follow this scheme.)

Omar takes as his theme the reality of our human condition. His verses state the human situation, that our life is brief; we have only a short time on Earth and will never return. Therefore we – that is, for him, men – should enjoy God's gifts while we may. These gifts are, above all, love of a beautiful woman and the power to forget our condition, if only temporarily, by wine.

It is remarkable that, in spite of the fanaticism of religious orthodoxy, Omar was able to write of such matters in a sceptical and rationalist vein. There is no evidence that he shared his thoughts with anyone or that he felt the need for such support. But some of his ideas – that wine is a symbol of God's gift to mortals, or that God needs human forgiveness rather than the other way round – seem lucky to have escaped causing the gravest offence.

> The dawn is here; arise my lovely one,
> Pour the wine, but slowly, and touch the lute,
> For those who are here will not stay long,
> While those departed never will return.

Omar Khayyam's algebra

Although al-Khwarizmi's book on algebra was already about 400 years old when Omar began his work, arithmetic and algebra were still not clearly distinguished. They were both designed to find the values of unknown numbers by relating them to known numbers. Omar made the formal distinction of defining algebra as the use of equations to find the unknown numbers by means of complete polynomials. (The word 'polynomial' refers to expressions that involve letters as symbols and that may involve more than one power of these letters.) He disagreed with the Greeks' refusal to recognise as numbers 'irrational' numbers (those that cannot be expressed as fractions, one like the square root of 2). His unique contribution, however, was to recognise 25 types of equations in place of the six kinds of al-Khwarizmi (see page 148). Fourteen of the 25 were connected to new procedures that he used for solving cubic (third-degree) equations. They involved new algorithms that called for the use of conic sections. These could be represented by quadratic equations, which stood for such geometrical figures as the circle, ellipse, parabola and hyperbola, or solid bodies such as the cube, dodecahedron and tetrahedron.

Suppose, for example, we had to find the value of x in an equation of the form

$$x^3 + ax = b$$

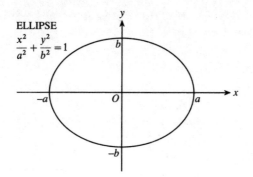

ELLIPSE

$$\frac{x^2}{a^2} + \frac{y^2}{b^2} = 1$$

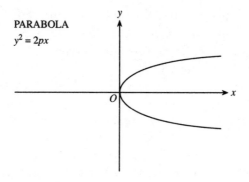

PARABOLA

$$y^2 = 2px$$

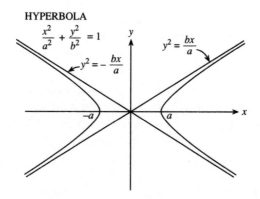

HYPERBOLA

$$\frac{x^2}{a^2} + \frac{y^2}{b^2} = 1$$

$$y^2 = \frac{bx}{a}$$

$$y^2 = -\frac{bx}{a}$$

where a and b are ordinary positive numbers such as 2 or 5 or 3. To solve this, we first write it in the form

$$x^3 + p^2x = p^2q$$
$$(\text{where } p^2 = a, q = \tfrac{b}{a})$$

Then we draw a circle where the coordinates x and y are related to each other in the following way:

$$x^2 + y^2 = qx$$

then the parabola $x^2 = py$.

Thus, we have created a simultaneous equation in two unknowns, x and y, in place of the original cubic. Omar could deal with these equations by drawing their graphs (the answers being given by their intersection), or by a more direct algebraic method. Where the graphs of the parabola and circle cut each other, both equations are true. The points are therefore also the solutions of the original cubic equation.

Suppose we illustrate this by solving a cubic equation using Omar Khayyam's method. We will also solve the conic section equations by an algebraic method. The equation used to illustrate the algorithm for this first type of cubic is the following:

$$x^3 + 4x = 16$$

We begin by putting this in the form:

$$x^3 + 2^2 . x = 2^2 . 4$$

(NB. The dot in this expression, and in those that follow, is a multiplication sign.)

This gives us the values of p as 2, a as 4 and q as 4. This tells us that the parabola and the circle have the equations

circle $x^2 + y^2 = 16$
parabola $x^2 = 2y$

We can solve these equations either by graphical means, or more simply as a problem in algebra. In both cases, we find that the value $x = 2$ is a solution. Thus, $x = 2$ satisfies the conditions of both the circle equation and the parabola equation. Our interest in it is that $x = 2$ also satisfies the original cubic.

To show why this solves the cubic, we do some elementary algebra:

The equation of the circle is $x^2 + y^2 = qx$
This can be written in the form $\frac{x}{y} = \frac{y}{(q-x)}$
The equation of the parabola is $x^2 = p.y$
This can be written as $\frac{p}{x} = \frac{x}{y}$
Multiply first equation by $\frac{x}{y}$: $\frac{x^2}{y^2} = \frac{x}{(q-x)}$
   But, by the second equation, $\frac{x^2}{y^2} = \frac{p^2}{x^2}$
   Therefore $\frac{p^2}{x^2} = \frac{x}{(q-x)}$

Geometric solution	Algebraic solution
Draw graphs of circle and parabola: 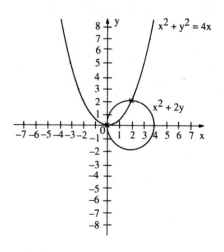	$x^2+y^2=4x$ $x^2=2y$ From the first equation $x^2-4x+4+y^2=4$ i.e. $(x-2)^2+y^2=4$. Now, for whole number solutions, $(x-2)^2$ and y^2 must be 0, 1 or 4, since these are the only whole number squares less than 4. By trying $y^2=1$ and $(x-2)^2=1$ we see easily that these are impossible. Therefore we have $(x-2)^2=0$ and $y^2=4$ or $(x-2)^2=4$ and $y^2=0$ i.e. $x=2$ and $y=\pm2$ or $x=4$ or 0 and $y=0$. We try the various possibilities in the second equation, and see that the only ones that work are $x=y=2$ and $x=y=0$
The circle intersects the parabola at the points $x=y=0$ and $x=y=2$, so these are the possible solutions of the twoequations.	The second cannot possibly solve the original cubic, but the first does; so the solution we want is $x=y=2$.

When we multiply both sides by x^2 and then by $(q-x)$, we see that this is the same as $x^3+p^2x=p^2q$. So a solution for this cubic equation must be a solution of the two quadratics, and vice versa.

Omar Khayyam gave routines for solving the other types of cubic equations. The general method he used was to put the equation in one of his standard forms. Then he selected two conic sections so that, when combined (in the fashion shown above), these could be reduced to the standard cubic for which he knew the solution.

Omar could solve other cubics, using another standard form: those of the form $x^3+a=bx$. This type can be solved by combining the parabola $x^2=y.\sqrt{b}$, and the hyperbola $x^2-\frac{a}{b}, 2x^2=y^2$. He specifically recognised that a cubic equation may have more than one root, although we know, of course, that he is interested only in the whole-number, positive solution. He understood that there could even be a negative solution. But he showed no interest in this, as negative numbers were impossible for him to deal with (even to conceptualise) and had no function in problem solving. This is strange since, in drawing graphs of conic sections, you must use both positive and negative numbers. You can actually see the negative roots. He just failed to make the connection. He also failed to recognise that if it is possible to divide both sides of the cubic by the unknown x, thus converting it to a quadratic, this meant that $x = 0$ was also a root.

Omar Khayyam and the binomial theorem

Sir Isaac Newton is honoured as the founder of modern mathematics because of his work on the calculus. His other key work in algebra was on the binomial theorem. This is where the sum of two numbers, say $(a + b)$, is raised to a power, say 1, 2, 3 or more. Newton discovered how to expand this kind of expression. The problem here is that, although $(a + b)$ to the power one is equal to $(a + b)$, when it is raised to a higher power than one there are intermediate terms as well. For example, $(a + b)^2$ equals $a^2 + 2ab + b^2$. Similarly, $(a + b)^3$ is equal to $a^3 + 3a^2.b + 3a.b^2 + b^3$. To the power four it is equal to $a^4 + 4.a^3.b + 6.a^2.b^2 + 4.a.b^3 + b^4$, and so on. As we increase the power, the number of terms also increases.

As we saw earlier, to establish the coefficients of the expanded polynomial, the Chinese mathematician Chu Shi-Chieh devised an algorithm. Omar Khayyam either borrowed it or reinvented it in his treatise *On the Difficulties of Arithmetic*. It is known in Europe as Pascal's triangle, and has been a matter for intense discussion because it reveals many properties of number series. It is also a useful tool for theories of chance and probability (see page 183).

Power of $(a + b)$	Coefficients of expansion
zero	1
1	1 1
2	1 2 1
3	1 3 3 1
4	1 4 6 4 1
5	1 5 10 10 5 1
6	1 6 15 20 15 6 1
7	1 7 21 35 35 21 7 1
etc.	etc.

The algorithm is simple. Each line after the second (1   1) is derived from the line above by adding two terms from left to right and placing the sum diagonally back one space in the next line. In the table above, the next line would be

$$1 \quad 8 \quad 28 \quad 56 \quad 70 \quad 56 \quad 28 \quad 8 \quad 1 \quad \text{etc.}$$

The Arabs' contribution

The Arabs' major contribution to science and mathematics was made during the golden years of Muslim supremacy. Their massive programme of translating into Arabic the scientific works of the Babylonians, Egyptians, Greeks, Indians and Chinese preserved what was known, making it available to Western scholars. It laid the basis for the Western scientific revolution of the 15th to 16th centuries.

The Arabs also developed new branches of mathematics, such as algebra and trigonometry. They laid the foundations of analytical geometry. They preserved Greek geometry and number theory from loss through disuse, translating a large part of the body of Greek mathematics – including, for example, Euclid's *Elements* and the *Arithmetics* of Gerasa and Diophantus. Six hundred years before Napier they set out the basic ideas behind the invention of logarithms.

It has been traditional in Christian Europe for some four centuries to denigrate the Arab contribution to mathematics. In other sciences, such as chemistry, botany, pharmacy and medicine, historians have been more objective in recognising the Arabs' achievement. But Arab originality in mathematics has been denied. It is pointed out that many who are credited with remarkable advances in 'Arab' mathematics in fact came from other groups, for example the Jews (Moses ben Maimon), the Persians (Omar Khayyam) or the Egyptians (al-Faradi). Exactly the same comment could be made about 'Greek' mathematicians. Two famous names may stand for those of dozens of 'colonials' who contributed. Archimedes was born and lived in Syracuse, in Sicily; Ptolemy was an Egyptian. They wrote in Greek, as Omar composed his scientific work in Arabic (only his verses were in his mother tongue, Persian). But as there is no quarrel with counting Archimedes and Ptolemy as Greek scientists, so there should be no conflict in accepting Omar and others as Arabs. They worked in Arab establishments, with Arab colleagues and Arab facilities. They wrote, thought and talked in Arabic.

In any case – and this is the main point – both Greek and Arab insights were derivative. A large proportion of their discoveries had already been made by the Chinese, Indians and Babylonians. Neither should it be forgotten that the brilliance of the Greeks as mathematicians was confined to a small number of thinkers who developed geometry as a branch of deductive logic. They failed totally to devise a suitable number notation. The 'Golden Age' of Greek mathematics consists of the works of Euclid, Archimedes, Ptolemy and Diophantus.

A proper comparison of Arab mathematics is therefore not with ancient Greece, but with Europe in the 7th to 15th centuries. It is a fact that, except for Spain (where the Arabs had a dominant influence), they had few rivals; Europeans who excelled in mathematics were those who had studied in Arab institutes of learning: Pope Sylvester II, Leonardo of Pisa, Fibonacci and a few others.

The Arabs' three major contributions were as follows:

First, they invented and spread knowledge of the decimal system. They set up the place-value method of expressing numbers. They made it possible for those who followed them in mathematics to understand numbers as an abstract system.

Second, they made it possible to see, in a decimal context, that fractions and integers as well as all other kinds of numbers can be brought under the same general rules if we define them appropriately. In particular, they laid a basis for interpreting negative numbers and treating roots and powers as complementary.

Third, they demonstrated that various kinds of number systems are not only possible but are interchangeable. The same results are obtained whether we use the decimal or the sexagesimal system or the binary system. There are advantages and disadvantages in the use of any system, but each is a direct analogue of the other.

FRANCIS BACON AND NEW DIRECTIONS

Having a mind nimble and versatile enough to catch the resemblance of things . . . and at the same time steady enough to fix and distinguish their subtler differences; as being gifted by nature with desire to seek, patience to doubt, fondness to meditate, slowness to assert, readiness to reconsider, carefulness to dispose and set in order, and as being a man that neither affects what is new nor admires what is old, and hates every kind of imposture, (so) I thought that my mind had a kind of familiarity and relationship with truth.

Francis Bacon, De Interpretatione Naturae

. . . his educated confrères . . . apparently considered him . . . something of a crank . . .

Maurice Cranston

THE details of Bacon's public career are well known. Born in 1561, he was called to the Bar in 1582, entered Parliament in 1584, and was a favourite of the Earl of Essex until 1601 (when he prosecuted the Earl for treason). Later, he served as Attorney General and Lord Chancellor. In 1621 he was condemned for bribery and corruption, fined and dismissed from office. Five years later he died. Under his own name he published some 30 philosophical works, most of them in Latin, on such subjects as the nature of knowledge, the logic of science and the ideal state. It was his declared aim to reform, improve, reorganise and harmonise all branches of knowledge.

Aside from these facts, an extraordinary group of legends (which may or may not be true) has attached itself to Bacon's name. If they are to be believed (and their supporters produce evidence of varying degrees of probability), then his 'secret life' includes the following details. First, he was a product of the morganatic marriage of Queen Elizabeth I and the Earl of Leicester, and the Earl of Essex was his younger brother by the same parents. Second, in an attempt to revive English literature almost single-handedly, he wrote not only all the plays of Shakespeare, but also poetry and other pieces attributed to John Lyly, Edmund Spenser, Thomas Watson and others. The writers whose names he used were paid money for this use. Third, his plea of guilty at his trial for bribery was forced from him by James I, to cover up a number of fraudulent practices by the King and his favourite Buckingham. Fourth, he developed Freemasonry (an amalgam of the rituals of mediaeval trade guilds and ancient Egyptian and Persian numerology). He introduced this secret society into Britain. Lastly – and this was no secret – was an idea that is of particular interest to scientists and philosophers. This was Bacon's view that the development of the natural and physical sciences 'to the limit' is necessary. As he put it, in a ringing phrase, this 'ministers to the relief of man's estate'.

The sure basis of knowledge: experimental science

In *The Advancement of Learning* (1605), Bacon's major contribution, he set out the method for his 'Great Reconstruction' of all branches of human knowledge. His starting point was that the art of thinking itself had fallen on evil days, being possessed by arid scholasticism and a reliance on authority (such as that of Aristotle) rather than on experience. He said that this led to sterile argument instead of illumination, and that people were 'very apt to condemn truth, on account of the controversies raised about it'. His remedy, the only sure way (he claimed) to advance understanding, was inductive reasoning.

Bacon identified 'learning' with what he called philosophy, and philosophy with what is now called empirical science. He said that three ways existed for advancing our understanding of philosophy: two wrong and one right. The wrong ways, tried and found wanting in the past, were, first, groping in the dark, without a plan; second, following the authority of the ancients. The right way is to build up knowledge by testing it as we go, proceeding from experiment to experiment. The laws of nature must be learned: they cannot be deduced by logic from definitions, nor from books, but only by close study and experience of nature herself. It is, Bacon claimed, humanity's prime task in this world to master nature.

In Bacon's elaborate defence of scientific procedures, and of the knowledge that matters, he exposes 'the errors and vanities which have intervened in the studies of the learned'. He is preoccupied by the need to get rid of the degenerate Aristotelianism which was the foundation of scholastic philosophy in its many forms. The effects of this argumentation are that philosophy, which should enlighten the mind, 'degenerates into childish sophistry and ridiculous affectation'. In place of this dialectical discussion based on word-chopping, he wants a new basis or foundation for natural knowledge. This can only be increased, and applied, through continuing, widespread observation and experiment.

One of the most striking of all Bacon's ideas is the new position he assigns to metaphysics. Traditional metaphysics had concerned itself with the notions of being and non-being, of accident and essence, of substance and form. These and other such categories provided opportunities for interminable discussion. Logical consistency, not correspondence with reality, was the criterion of truth. Bacon rejected all such categories, redefining metaphysics itself as the basic science of material physics. In place of authority and logical analysis of propositions as the prime tests of metaphysical truth, he described three levels of natural inquiry: the collection of scientific data; the investigation of particular natures and causes; the clarification of the fundamental forms of things. His other striking innovation is a sharp separation of human truth from the dogmas of revealed religion. He allows reason only a limited function in deriving practical consequences from the divine mysteries.

None the less, Bacon is careful not to attack religion. The truths of religion are 'placed beyond, and exempted from, the examination of reason'. It is futile to seek knowledge of nature in divinity, or of sacred theology in natural inquiry. This separation of knowledge from faith is in accord with his view that human knowledge (based as it is on experience) is limited to sensible, material and finite things. He deliberately limits himself to the analysis of experimental knowledge, and never discusses the nature of religion or, in detail, how faith differs from experimental knowledge. He seems to assume that there would be no profit in doing so.

In his *Novum Organum* (1620), Bacon further expounded his grand programme for the reconstruction of knowledge. The book's very title ('New Method of Reasoning') gives notice that Bacon is opposed to the method that holds the high ground, namely, Aristotle's deductive *Organon* of logical analysis. Bacon's new method is inductive: he argues from particular cases to a general formula. Aristotelianism, and the Christian scholasticism derived from it, proceeded by definition and deduction. Bacon's method is to test assertions of the connections between things by checking whether these connections really exist. That is, he states or provides the proposed or declared connections and then verifies that the phenomenon then appears, without exception. On each repetition of the experiment, the phenomenon again appears.

In *Novum Organum*, Bacon enumerates the hindrance to knowledge: the 'idols' that occupy human minds to the exclusion of truth. These 'idols' are such things as false opinions, dogmas, superstitions and errors. He describes four kinds of idols in particular, the four main sources of error which interfere with the acquisition of correct knowledge by colouring and distorting our understanding of the nature of things. They are:

(i) *idols of the tribe*. These are errors that result from human nature directly. Our thinking is improperly affected by our personality, will and passions. In the normal course of events, the mind receives impressions from objects. But we are 'preoccupied' by opinions that are derived from earlier experience. What a person prefers is what he or she more readily believes. We reject ideas and new experience because of the way our passions influence and corrupt our understanding. The mind receives those things that impress it. An agreeable opinion leads us to accept only those facts that seem to support it. We overlook, or easily forget, anything that seems to suggest the contrary.

(ii) *idols of the cave*. These are errors that arise from our individual nature, education and environment. We are all cut off, or shut up, in a cave made from our individual peculiarities. Each of us becomes obsessed with a certain method, and with certain concepts or presuppositions. We proceed to compose what we believe to be a complete body of knowledge based on these premises. Bacon cites Aristotle as an example of this, picturing him as constructing a whole philosophy by deduction from a few scraps of experience. Similar mistakes are made by those who choose a particular finding as being 'by the wise', ignoring what is affirmed by opponents or thinkers of other times. Again, some investigators consider Nature in its totality, others break it up. However, the most fruitful scientific procedure is to combine the two techniques of analysis and synthesis in a single 'analysis of experience'.

(iii) *idols of the market-place*. These are the most troublesome. They are errors that arise by virtue of human intercourse and from the peculiarities of language. Words do not always convey the correct meaning, so they often serve to weaken knowledge. Words are misused when they refer to things that do not exist in reality, but only as confused abstractions. When a word needs to be altered in sense, or a new one employed, the innovator meets with opposition. People believe that they control words, 'yet certain it is that words, as a Tartar's bow, do shoot back upon the understanding of the wisest, and mightily entangle and pervert the judgement'.

These three classes of idols are inherent in human nature. They can be controlled but not eradicated. A fourth type, *idols of the theatre*, is not inherent in nature. They may be prevented from entering the mind, or expelled from it. They arise from teachings based on philosophical systems. Their common feature is that they are elaborate representations of nature based on a minimum foundation of observation, experiment, and natural history.

The main significance of this complex classification of human error is that it seems designed to establish that no human mind is a repository of divine truth, absolute and universal. If scientific truth is to be based on experience, we need to be on our guard against influences that corrupt our understanding of the data of experience. But Bacon is careful not to highlight the divisive features that separate humans, and does not elaborate. His intention seems to have been to lay the basis for the study of individual differences, by pointing out the sources and the fact of variety in human thinking about the nature of reality.

In the *Novum Organum*, Bacon sets out his programme for a logic that is occupied with the results of accurate observation and experiment. He describes how the scientific method of inquiry, 'the true method of experience', first 'lights the candle [hypothesis], and then by means of the candle shows the way [arranges the experiment], commencing as it does with experience duly ordered and digested, not bungling nor erratic, and from it deducing axioms, and from established axioms again new experiments'. He states that the aim of this method is to discover the fixed and essential laws that describe the operation of material bodies. This, he says, is what philosophers really mean when they talk of 'forms':

In reality, when we speak of forms we mean nothing less than those laws and regulations of simple action which arrange and constitute simple nature . . . the form of heat or the law of light . . . For although nothing exists in nature except individual bodies exhibiting clear individual effects according to particular laws; yet, in each branch of learning, those very laws – their investigation, discovery and development – are the foundation both of theory and practice.

Bacon's doctrine of forms is based on 'naturalism', which interprets reality as operating in terms of the principle of determinism. The Universe is not, as some suggest, an organism with a soul. It is a kind of machine. When science has discovered the forms of things, the world will be merely the raw material from which human beings can proceed to construct whatever utopia they choose. It is his expectation that these theories will be hailed, by people of faith and learning, as a sound philosophy.

Bacon and God

Christian Aristotelians identified the First Cause with the God of Christian revelation. Christian Platonists believed in a mystical illumination which enabled human beings to see truth directly. Bacon takes up a third position. He accepts a Creator but rejects sectarianism. The world had to wait three centuries for the word to be coined (by Thomas Henry Huxley), but 'agnosticism' would probably be the best description of Bacon's religious views – provided that we could define agnosticism to mean that we know, as a balance of probabilities, that God exists, but, except that He created the universe, we know nothing further about Him. (This differs radically from

Huxley's position, which works on the basis that we can be 'fairly sure' that God does not exist but are not prepared to say so – an attitude once characterised as 'shame-faced materialism'.)

Bacon refuses to construe his doctrine in Platonic or in Aristotelian form. He contends that we may properly philosophise and, through our philosophy, come to know the goodness and power of God as set forth in His created works. But Bacon is unable to see in the human creature any natural faculty that enables us to discern the nature of God. Wisdom respecting the supreme being cannot be attained by reason and sense but only through revelation. He does not single out any particular body of beliefs as acceptable as God's revelation to humankind.

This separation of natural philosophy from theology is fundamental to Bacon's philosophy of nature, humanity and God:

It is true that the contemplation of the creatures of God hath for end (as to the natures of the creatures themselves) knowledge, but as to the nature of God, not knowledge, but wonder . . . Therefore attend His will as Himself openeth it, and give unto faith that which unto faith belongeth; for more worthy is it to believe than to think or know.

Bacon and the invention of word processing

In *The Advancement of Learning*, Bacon discourses on the use of ciphers and codes, giving some examples. He describes what he calls the 'bilateral alphabet code'. This consists of what he calls two alphabets, each of which comprises one letter, A or B. In fact he is describing the binary system, using A and B where we would nowadays use 0 and 1.

The code, as used for the English alphabet, consists of translating each individual letters of the message into As and Bs according to a certain scheme (which Bacon, envisaging a secret code, said should be known only to the sender and recipient of the message). We have five places to fill either with an A or B, that is, five choices to make between the two letters. In other words, we could accommodate 2 to the power 5, that is, 32 choices. This is enough since we only need 26 for letters of the alphabet. Bacon coded the letters as follows:

A	AAAAA	G	AABBA	M	ABABB	S	BAAAB
B	AAAAB	H	AABBB	N	ABBAA	T	BAABA
C	AAABA	I	ABAAA	O	ABBAB	U	omit
D	AAABB	J	omit	P	ABBBA	V	BAABB
E	AABAA	K	ABAAB	Q	ABBBB	W	BABAA
F	AABAB	L	ABABA	R	BAAAA	X	BABAB
		Y	BABBA	Z	BABBB		

Some centuries have passed since the 'fine summer's day' in Paris when Bacon devised this code. That was the moment when the basic idea of word processing was born. Many support systems – typewriters, cathode-ray tubes, current electricity, telegraph and the transistor – were still in the womb of time; the binary number system had still to be imported from China. But Bacon contributed the basic idea from which all modern systems were developed. The American Standard Code for Information Interchange (ASCII) was accepted in 1966, at least temporarily, for transmitting messages by telegraph. In this, there are 2 to the power 7 (or 128) different alphabetic, numeric and control code numbers corresponding both to characters and to controls such as 'backspace', 'carriage return', and so on. These code numbers, expressed in binary, do the same job as Bacon's five-letter alphabetical formulas. The letter G, for example, represented in Bacon's code by AABBA, is in ASCII 0100 0111 (decimal 71). The ASCII code is used in modern word-processing computers. In practice, the computer translates all keyboard letters, spaces, (Arabic) numbers and keyboard controls when struck, into their ASCII equivalents for internal work, and for storage in the memory. It converts these numbers back from ASCII to letters, spaces and control commands, which are then displayed on the screen or printed out in the same form as the original document. The main ASCII codes are shown in the table.

Code numbers (given in decimal)	Typewriter characters											
49 to 57	1	2	3	4	5	6	7	8	9			
58 to 64	:	;	<	=	>	?	@					
65 to 78	A	B	C	D	E	F	G	H	I	J	K	L	M
79 to 90	N	O	P	Q	R	S	T	U	V	W	X	Y	Z
91 to 96	[	\	]	^	–	'						
97 to 109	a	b	c	d	e	f	g	h	i	j	k	l	m
110 to 122	n	o	p	q	r	s	t	u	v	w	x	y	z

Bacon's contribution

Bacon is regarded by those acquainted with the history of science and philosophy as providing the main key to the modern world outlook. His criticism of scholastic philosophy and the Greek dialectical heritage brought a new spirit into the arguments about the origin of things, and about the way truth should be pursued, with some hope of a positive outcome. Although he never trained as a scientist (or perhaps because of it) he was able to bring the light of reason to bear on the errors and superstitions of the learned.

Bacon's most important writings were centred on 'method'. He was able to cut through the detail of the new science and isolate the leading characteristic – that of scientific inquiry, as opposed to philosophical discussion and rhetoric. The method of science is *induction* from observation and experiment. 'The Book of Nature' provides the scientist with both the method and the data which eventuate in truth.

Bacon is silent on a great number of questions that were bedevilled by controversy – and, in his refusal to become involved in traditional metaphysics and theology, his silence is more eloquent than argument piled on argument by the disputants. He turned away from verbal argument, and provided a new role-model for the scientific era he inaugurated. It is true that he made no contribution to number as such. But what he contributes in terms of attitude and basic assumptions enlightens our understanding of the nature of numbers, and releases us from thraldom to the idealism of the Platonic school and the Roman scholasticism which dominated thought in his day. He is the harbinger of the new era.

JOHN NAPIER: THE RATIONALISATION OF ARITHMETIC

Let it be your Majesty's continuall study to reform the universall enormities of your country, and first to begin at your Majesty's own house, familie and court, and purge the same of all suspicion of Papists and Atheists and Neutrals . . .
from Napier's dedication of his Plaine Discoverie *to King James VI of Scotland*

IF John Napier (1550–1617) had been asked, he would certainly have said that his major work for humanity was his campaign against Roman Catholic supremacy in Scotland and later in England. A fanatical Scottish Presbyterian, and follower of John Knox, he spent much of his adult life inveighing against such Catholic European monarchs as Mary Queen of Scots, Charles IX of France and Philip II of Spain. For 27 years, from the age of 16 until he was 43, he worked on a book, *A Plaine Discoverie of the Whole Revelation of Saint John*, whose principal purpose was to identify the Pope (any Pope, or maybe all Popes) as the Antichrist. As the quotation that heads this chapter shows, Napier, so far from fawning on his royal patron, gave the king a characteristically uncompromising homily.

Today, such frenzy seems more pathetic than significant. Napier's important contributions are not to religion but to mathematics. Yet, his *Miraculous Canon of Logarithms* and the calculating device nicknamed 'Napier's bones' were the fruits of his leisure: his calculations for the book on logarithms alone took 20 years. His mathematical genius and insight are such that we can only regret that he never made number-study his main work. The world of knowledge is probably much the poorer for the fact that he spent so much time on anti-Popish propaganda.

However, there is a link between the two sides of Napier's life. His fanatical dedication to the Protestant cause, his obsessive persistence and his devotion to detail are all exactly matched in his approach to mathematics. His ambition was to free the minds of those bedevilled by numbers, as one might free the soul from the trappings of Popery – to cleanse arithmetic in the same way as Knox or Calvin had cleansed Christianity. In fact, he wanted to abolish arithmetic altogether, as it was currently practised, and to replace it with a rational system so simple that anyone – a child, or even a machine – could do it.

One example will suffice to show how Napier's solutions differed from those in force at the time. Although Arabic numerals had begun to replace Roman numerals by his time, there was still no standard way of writing fractions. Napier pioneered the use of the decimal point. (In fact he was, in this, ahead of his time: it was not until two centuries later that his use of the full stop – or occasionally the comma, as used in continental Europe – to separate the decimal fraction from the whole number, was universally adopted.) To illustrate the difference, here are three contemporary ways of showing the same fraction:

Simon Stevin (1585):
$$139 \quad \overset{,}{2} \quad \overset{,,}{5} \quad \overset{,,,}{6} \quad \overset{,,,,}{8}$$

Henry Briggs (1619): $139 \quad \frac{2568}{10000}$

$$139 \quad 2568$$

Napier (1617):   139.2568 or 139, 2568

'Napier's bones'

In his book *Rabdologiae* (1617), written in Latin, Napier comes closest to the interests of ordinary people. He is probably the first, as Babbage is the greatest, of the public benefactors who took up the study of mathematics. In *Rabdologiae* he describes his famous 'bones', or 'divining rods' as he called them. They are a simple application of the 'galley' method for multiplying, as used in his time. They take most of the pain out of multiplication and division, and can also be used to find roots and powers. All you need to know are the Arabic numbers 0 to 9, what Arabic number notation means (see opposite), and how to add and subtract. The 'bones' replace tables and prefigure Napier's invention of logarithms.

The 'bones' can easily be made using strips of paper or pieces of wood. Ten strips about 5 inches long and half an inch wide are needed. You draw lines and write numbers on these strips in the manner shown below. Each strip serves as a line from the multiplication table. The 'bones' can also be used (in reverse) to do division.

For example, to multiply 643 by 249, we begin by removing the rods for 6, 4 and 3 from the box of 'bones'. We also remove the end rod, which serves as a marker. On this are written the Roman numbers I to IX. We lay the rods side by side, as shown.

1	2	3	4	5	6	7	8	9	marker
1	2	3	4	5	6	7	8	9	I
2	4	6	8	10	12	14	16	18	II
3	6	9	12	15	18	21	24	27	III
4	8	12	16	20	24	28	32	36	IV
5	10	15	20	25	30	35	40	45	V
6	12	18	24	30	36	42	48	54	VI
7	14	21	28	35	42	49	56	63	VII
8	16	24	32	40	48	56	64	72	VIII
9	18	27	36	45	54	63	72	81	IX

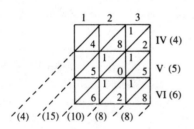

$$4 = 4 \qquad 4+1 = \boxed{5}$$
$$5 + 1 + 8 + 1 = 15 \qquad 5+1 = 6$$
$$6 + 1 + 0 + 1 + 2 = 10 \qquad 0 = 0$$
$$2 + 1 + 5 = 8 \qquad 8 = 8$$
$$8 = 8 \qquad 8 = 8$$

IV (4) V (5) VI (6)

(4) (15) (10) (8) (8)

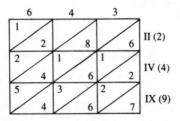

II (2) IV (4) IX (9)

Modern method	Napier's bones
643 × 249	643 × 249
5 787	9 × 643 = 5 787
25 72	40 × 643 = 25 720
128 6	200 × 643 = 128 600
160 107	249 × 643 = 160 107

Before proceeding, it is important to remember that 249 is Arabic number shorthand for $200 + 40 + 9$ and that this in turn is shorthand for $(2 \times 100) + (4 \times 10) + (9 \times 1)$. To multiply 643 by 249, we first copy down the result of multiplying 643 by 9 units, then by 4 tens and then by 2 hundreds. Then we add the results. (After using the rods two or three times, this way of working becomes automatic.) We look along line IX of the Roman marker, from right to left. We see that nine threes are 27 units; nine fours are 36 tens, and nine sixes are 54 hundreds. This means that we must carry a 2 into the tens place and a 3 into the hundreds place. The result of the addition is 5787. We write this down. Now, looking along line IV (knowing that this stands for 40), we write down 25,720. Then the II (meaning 200) gives us 128,600.

The advantage of Napier's device is that we can actually see the 'carries'. To multiply 643 by 9, we mentally slide the 2 in line IX down its diagonal to make 8 $(2 + 6)$ for the first partial result; we slide 3 down its diagonal to make 7 $(3 + 4)$. We can then read off the result, 5787.

Next, to multiply 643 by 40 we go through precisely the same operations. We look along the line marked IV. We see that there are two carries here, both of 1. The answer is 25,720 (remembering we are multiplying by 40, not 4). Lastly, we multiply by 200. Here we find the Roman number II on the marker. There are no carries this time. The answer is 128,600.

To complete the operation, we add the three 'partial results' together: 643 multiplied by 249 is equal to the sum of 128,600, 25,720 and 5787. The answer is 160,107. This can be checked by multiplying 643 by 249 on paper, or by using the 'bones' to find 249 times 643.

Division is done in much the same way. The difference is that we look in the 'bones' for the result: we do not start from the marker; we finish there. To divide, say, 160,107 by 249, using the 'bones', we do all the steps in reverse. We look for a number equal to, or just less than, 160,107. We find it in line II. We subtract this number, 128,600 from 160,107 to leave 31,507. Then, we look for a number equal to, or just less than, 31,507 (line IV) . . . and so on. The advantages of the method are, first, that we learn to dispense with number tables; second, that we think about numbers in a new way, not taking them for granted; and third, that we cease to fuss about 'carries', these being done quite automatically.

This may not seem a tremendous advance. However, we need to remember that we are sophisticated persons, growing up in a civilisation dominated by numbers and letters. Without knowing it, we live on mental capital accumulated by other people. As in the case of Egyptian fractions (see page 45), we need to compare the simplicity of Napier's methods, not with what we know now, but with what was the practice in his own time. Again, an illustration will put things in perspective. Fifty years after Napier, Samuel Pepys needed some basic mathematical skills in his job as Secretary of the Navy, calculating victuals, pay and other such matters. To do even the simplest calculation, for example to multiply two numbers less than 10 and greater than 5 (let us say 8×7, as below), he had to go through the following mental contortions:

To multiply 8 by 7	
First set your digits one over the other	$\begin{matrix} 8 \\ 7 \end{matrix}$
Then draw a cross	$\begin{matrix} 8 \\ 7 \end{matrix} \times$
Then write the difference of each from ten	$\begin{matrix} 8 \\ 7 \end{matrix} \times \begin{matrix} 2 \\ 3 \end{matrix}$
Now multiply these differences	$2 \times 3 = 6$
Now write the product (6) in the units place	6
Now subtract across a diagonal ($8 - 3$ say, or $7 - 2$)	5
Write this difference in the tens place	50
Add the six	56
Seven times eight is fifty-six.	

The invention of logarithms

Napier realised that all numbers could be thought of as being one continuous series. The importance of this is that the same rules apply throughout to whole numbers, to fractions and to mixed numbers. We can think of this series as an uncountable number of points very close together on a line. Whole numbers have always seemed to humans to be of particular significance, standing out like milestones on a road. But this is an illusion. Napier saw that there is really nothing special about whole numbers, except maybe that they are easy to remember.

This new concept of number was the basis of the invention of logarithms. Napier discovered the relation between two well-known series of numbers, the addition or arithmetic series, and the multiplication or geometric series. His insight was that this relation was true for all numbers, whether whole numbers, fractions or mixed numbers (whole numbers plus a fraction). To state this relation as simply as possible, he discovered that one series could be written in terms of the other: a geometric series could be written as an arithmetic series, and vice-versa.

Using the 2s series as an example, we can generate the two kinds of progression as follows:

Arithmetic	0	1	2	3	4	5	6	7	8	9	10	
Geometric	1	2	4	8	16	32	64	128	256	512	1024	

The relationship between the two series is as follows:

2 to the power	0	1	2	3	4	5	6	7	8	9	10	
is equal to	1	2	4	8	16	32	64	128	256	512	1024	

These series illustrate the binary system, used in computer work. The multiplier (or base) of the geometric series is 2. There are any number of such series. We can use 2 as our base, or any other number we like. The chosen number is the 'root', or 'base', by means of which we can generate any other geometric series. The two series, arithmetic and geometric, are directly linked by our choice of base: 2 in this case. When we raise the base (2) to the different powers set out in the arithmetic (addition) series, we obtain the geometric (power) series. We normally write the power above the base, as follows:

Power	0	1	2	3	4	5	6	7	8	9	10
Base	2	2	2	2	2	2	2	2	2	2	2
Equals	1	2	4	8	16	32	64	128	256	512	1024

This perception was Napier's starting point. As we said, he saw that every number, whether a fraction or a mixed or whole number, can be expressed as a power of another number, which we now call the base. It is not just that they are equal. Fundamentally, it means that we have two ways of writing the same numbers. There is a given number, say 8, there is the base, say 2, and there is the power to which the base must be raised to equal the given number, in this case 3. To the power, Napier gave the name the 'logarithm' of the number. We now have three ways of conveying exactly the same information: 'two times two times two is eight'; 'two to the power three is eight'; 'logarithm eight to the base 2 equals 3'.

The next stage is to see that when three numbers are linked (given number, base and logarithm), if one changes, so does one other. If the problem is to multiply two numbers, we add the logarithms; if dividing, we subtract them. The process can be illustrated by taking two numbers from the binary series (base 2). Suppose we want to multiply 64 by 4. We begin by expressing them as powers of 2. Then, to multiply the numbers, we add their logarithms. So:

$$64 \times 4 = 2^6 \times 2^2 = 2^{8(=6+2)} = 256.$$

To divide 64 by 4, instead of adding the logarithms we subtract them:

$$\tfrac{64}{4} = \tfrac{2^6}{2^2} = 2^{4(=6-2)} = 16.$$

These examples are really too easy to calculate by logarithms. But working with larger numbers is just as easy. For example, to multiply 23.9657 by 19.2739, you look up the logarithms of the two numbers, and add the logarithms together. Then you look up that result in a table of antilogarithms, and find the answer. (In this example, the logarithms are to base 10.)

---

logarithm 23.9657 = 1.379590
logarithm 19.2739 = 1.234949
Sum of logarithms= 2.614539
Antilogarithm　　= 461.7299
Therefore 23.9657 × 19.2739 = 461.7299

Calculating logarithms

In retrospect, the easy part might seem to have been working out the principle of logarithms: that is, that there is a relationship between number, power and base, and that the principle applies whatever the number base. The base can be a whole number, a fraction, a mixed number, and it can be in any scale of notation, binary, octal, decimal, hexadecimal, sexagesimal or any other. The hard part, one might imagine, would be calculating the actual logarithms. Tables of logarithms and antilogarithms might make calculation as simple as ABC – but what about the labour of producing those tables in the first place?

Napier tried many ways of calculating powers to bases. For example, he would calculate numbers such as 2 to the power 10,000 (that is 2 multiplied by itself 10,000 times). He would then count the number of places in the answer. This comes out to 19,950,583,591, . . . to 3011 places. Subtract 1 from 3011 and you get the logarithm of 2, correct to four places of decimals. (The characteristic (whole number part) is 0, the mantissa (decimal part) is .3010, so the logarithm is: logarithm 2 to base 10 = 0.3010.)

Napier, who worked usually to seven places of decimals, soon decided that there was no future in this particular procedure (that is, continuous multiplication of two by itself). He set out to find other methods, and experimented with some half dozen, almost as tedious, before hitting on a solution which was as quick as it was accurate. To understand it, we must recall two different kinds of average, the arithmetic mean and the geometric mean. The one we use most, the ordinary 'average' of everyday affairs, is the arithmetic mean. The arithmetic mean of any two numbers (say 4 and 16) is found by adding them and dividing the answer by 2. (Thus 4 and 16 makes 20; divide by two, the answer is 10. The arithmetic mean of 4 and 16 is 10.) The geometric mean is used for scientific purposes. To find it, you multiply the two numbers (say 4 and 16, again) and take the square root of the result. (Thus, 4×16 is 64; the square root of 64 is 8; 8 is the geometric mean of 4 and 16.)

Remembering that a logarithm is a power, we can state the rule for any two numbers whose logarithms we know. The logarithm is the arithmetic mean of the original two logarithms; the number is their geometric mean. For example (working in base 10), let us take the numbers 100 and 1000. To what power do we have to raise the base 10 to get the number 100? What power for the number 1000? Answers: to the first question, 2; to the second, 3. The rule says: the unknown number is the geometric mean of 100 and 1000 (that is, the square root of 100×1000, that is, 316.226677). The logarithm is the arithmetic mean of 2 and 3 ($2+3$ divided by $2=2.5$). In other words, the logarithm to base 10 of 316.226677 is 2.5.

This was Napier's starting point, the discovery that unlocked the whole process. He was now able to use the process of recursion (see the table on page 173). He was able to draw out all sorts of conclusions, using this principle:

> Start with any two numbers and their known logarithms.
> For a new and unknown logarithm, work out the arithmetic mean of the two earlier logarithms.
> For the unknown number, work out the geometric mean of the numbers.

We can use this to plot all the numbers from 100 to 1000. For example, taking up the argument where we left it, the two numbers are now 100 and 316.226677. Their geometric mean is 177.827641. The arithmetic mean of their two logarithms (2 and 2.5) is 2.25. Therefore the logarithm of 177.827641 to the base 10 is 2.25. The recursive method is shown in the table.

This process combines the methods of interpolation and recursion. Interpolation is where we start with two points or numbers and find the number between them that satisfies the problem (in this case, the geometric

Numbers			Logarithms	
Step 1	100	1000	2	3
Means	GM = 316.227		AM = 2.5	
Step 2	100	316.2267	2	2.5
Means	GM = 177.828		AM = 2.25	
Step 3	100	177.8276	2	2.25
Means	GM = 133.352		AM = 2.125	

etc. (GM = geometric mean; AM = arithmetic mean)

and the arithmetic means). The method of recursion is where we go back with fresh values, and repeat the calculation with the numbers we have obtained in the previous step.

Napier's work was immediately accepted. Astronomers, ship's captains, scientists and engineers were grateful for being saved countless laborious hours of calculation. When James VI of Scotland (who also became James I of England) went to Norway in 1590 to collect his bride-to-be, Anne, his medical attendant was Doctor John Craig, a friend of Napier's. The boat landed near Hven in Norway. The astronomer Tycho Brahe had been observing the stars there for decades. He had data for hundreds of observations of the planets at different times of the year. Craig told him of Napier's logarithms. Kepler, Brahe's assistant (who inherited his observations) used Napier's logarithms on Brahe's data to establish his three laws of planetary motion. The work of Brahe and Kepler laid the foundation for Isaac Newton's revolution in the exact sciences (see Chapter 13). All this, and most of modern science, would have been impossible, or at least put back in time, without Napier's logarithms.

Napier's logarithms went on being used until the advent of electric calculators and electronic computers made them unnecessary. Until the late 1960s, all secondary-school children studying mathematics would have been familiar with (if not enamoured of) their books of tables, which were an indispensable tool. Furthermore, calculators and computers themselves owe much to Napier: his 'bones' are one of the first modern mechanical aids to computation. The recursive principle that he developed is one of the most basic ideas in programming. Using the recursive algorithm, we can now do the calculations on which he spent 20 years, in as many minutes. This is one way of measuring his achievement.

1 2 3 4 5 6 7 8 9 10 11 12 13 14 15 16 17 18

THE NEWTONIAN REVOLUTION: THE MARRIAGE
OF CRAFTSMANSHIP AND SCHOLARSHIP

It is useless, my son. I have read Aristotle through twice and have found nothing about spots on the Sun. There are no spots on the Sun. They arise either from imperfections of your telescope or from defects in your eyes.

17th-century Jesuit professor, as reported by Kircher

Surely you mistake Cambridge. Wee are situated in as dark a corner of ye land (in terms of knowledge of world affairs) as can well be desired.
Roger Cotes, first Plumian Professor of Astronomy and Natural Philosophy,
Cambridge, to his uncle, 1707

THROUGHOUT their history, the English have periodically experienced vast changes in government, in the economy and in social and religious affairs, with the whole face of society radically transformed. At other times, sometimes lasting for centuries, changes have been limited to minor rearguard operations. Whenever innovations have been accepted by the ruling group, they have been imposed, sometimes with great cruelty, on the majority.

The Tudor monarchs, especially Henry VIII, carried through the most massive changes in religion, property relations, education and social affairs since the Norman Conquest. To engage the energies of 'reformers' as agents of change, the drive was redirected from the bottom line of profit and loss to religious and other core values. Sequestration of the monasteries (lands, endowments, buildings and schools), dispossession of peasant farmers, enclosure of common land, the transformation of arable land into sheep pastures – these activities had the effect of projecting a whole generation of commoners, without notice, directly from what they could look back on as their golden age into what was undoubtedly the grim age of iron (in which many ordinary people still live).

In the normal course of affairs, the new social system – the new order of capitalist enterprise – brought the English aristocracy and the new owner-class into conflict with the Catholic powers, in the first place with Spain and the Vatican. Between the 1520s and 1580s Spain was establishing and expanding its colonial empire in Central and South America, shipping vast quantities of treasure home from the New World. Spanish ships were constantly harassed by English pirates whose activities were franchised by Queen Elizabeth.

It was not yet open war, but the Spaniards were preparing an invasion, and in May 1588 set off from Lisbon with 130 ships and 8000 sailors. The English navy at this time numbered 16,000 men and 197 ships. The ships were small, low-lying, manoeuvrable and lightly armed; the Spanish ships, by contrast, were large, heavily armed and clumsy. In three encounters in the English Channel the English ships harassed the enemy. Their light armament did little damage, but they took advantage of their greater manoeuvrability and used fireships to disorganise the Spaniards, who were also hampered by gales. The Spanish fleet sailed home in disarray with only 76 ships. The English lost no ships and fewer than 100 men. One immediate result was that half the young gentlemen of England decided to go to sea.

Navigation in Elizabethan times

The upsurge of interest in sea travel gives us a case study, one of dozens, for the absolute backwardness of science at this time. Entrepreneurial and, in this case, military and exploratory enterprise was hindered by the primitive state of knowledge of the natural world, let alone of mechanisms for putting that knowledge to constructive use. Sixteenth-century navigation (whether for exploration, trade, piracy or war) was a practical skill, a pragmatic affair with hardly any intellectual rationale. Most voyages were local in character, usually carried out within sight of land. The compass was in common use but nothing much was known about how it worked or about the corrections that had to be made in using it. (It was not until 1600 that William Gilbert, after years of study and experiment, published the first-ever book on magnetism and the compass, stating the correct view that the whole Earth behaved like an enormous magnet.)

Most serious of all, there were few reliable maps, especially for long voyages. Columbus' mistake in identifying America as India indicates the primitive level of geographical knowledge at the time. Professional navigators were appointed for journeys to unknown lands on the basis that they had been there before. They provided themselves with 'rutters' (from the French *routiers*) similar to the 'routes' provided today for motorists by national touring offices. These suggested the best possible route, warned of dangers, listed prominent and visible landmarks, gave distances between inhabited places, and so on. The rutter was the navigator's personal property, his main professional asset.

For shorter voyages, such as across the English Channel, the navigator would give oral directions from memory. Sailors would normally be illiterate. Only the captain and navigator could read the charts, set a course by the compass, measure the ship's speed and note distances in the log. The other main navigational aid was a timing device (the hour-glass), which might measure one- or two-hour stints, two of the latter in sequence counting as a 'watch'.

Sailing could be accomplished using only 'dead reckoning' (that is, by compass and hour-glass). An imaginary line of longitude would be followed, the ship's movement being timed by hour-glass. The speed of the ship could be estimated by means of a special device (also known as the 'log') thrown overboard, the distance travelled each day being noted.

The measurement of latitude was a relatively simple matter: there were several instruments available that could be focused on celestial bodies (the angle of the Sun to the horizontal at noon being the most obvious). This observation would provide the angle from which the latitude could be calculated. The quadrant and theodolite were similar instruments. But sailing to a given destination was at best a hit-and-miss affair, as there was no accurate way of determining longitude. Mariners were reduced to hugging the coastline, following their course on a rutter, or (on longer voyages) literally launching themselves into the unknown, putting their trust equally in God, rudimentary celestial observation and a vast mass of folklore and travellers' tales.

For English mariners at least, mathematics was a crucial problem. Continental sailors had access to printed scientific materials in their mother tongue, very often written by practical seamen. But the English had nothing similar. A few texts existed, written in Latin by scholars for other scholars, but they were of little use to illiterate seamen, or even literate ones. As late as the 1650s, when Samuel Pepys, as Secretary of the Navy, was working on the problem of providing technical instruction for mariners, he was thwarted by the total lack of teachers who combined practical experience in seamanship with knowledge of Latin. (Why Latin? You may well ask.) Boys anxious to follow a naval career (that is, to become officers) had to study at home, directed by their fathers or older brothers, or be instructed by private tutors or mathematics 'practitioners' in the capital. These people were usually scientific instrument-makers or experienced, retired naval officers or mariners. They would give practical instruction in geometry, astronomy, surveying, gunnery and horology.

The content of one such course is set out in Andrew Wakerly's book *The*

Mariner's Compass Rectified, with the Use of all Instruments of Navigation (1633). Wakerly was a maker of nautical instruments whose courses were held annually for over 30 years near Cherry Garden Stairs in London, across the Thames from Wapping Old Stairs. He died during the London Plague in 1665. His course contained the following elements (some words are given in modern form, in square brackets):

Arithmetic. In Whole numbers and Fractions, lineally, superficially [that is, using areas], cubically and Instrumentally performed: the Extraction of Rootes, as the Square, Cube, Squared square and Cubed cube. Decimal Arithmetic and Astronomical Fractions whereby are calculated the motion of ye Celestiall Bodies.

Geometry. By Demonstration and Practice, as to Measure the superficial [surface] quantity of Board, Glasse, Land etc. The reduction of Plotts or Mapps to any assigned proporcion: The taking of Heights, Distances and Profundities.

Trigonometry. Or the Doctrine of Triangles, Right-lined or Spherical, with ye Grounds and Demonstrations to every Case proved by the Tables of Artificial Sines, Tangents and Logarithms.

Gunnery. The Cheefe grounds and principles belonging to ye Art, Geometrically, Arithmetically and Instrumentally performed, likewise ye finding ye true Weight of any Peece [gun] by the Dimensions only and to give him his true allowance of powder and shott.

Gauging. Or ye speady Measuring of all manner of Vessels, either Wine, Oyle, Hony or Beere. Likewise ye finding the True Content of a Brewer's Tun, of a Pond, of a Ship's Hold, or of a River etc. Instrumentally and Cubically performed.

Instruments. Demonstrating and Teaching ye use of all Sorts, either for Sea or Land, for Operation or Observacion, ye Sphere and both Globes, Celestiall and Terrestriall.

Horologiography (*time-reckoning*). Or ye Art of Dyalling according to any Plaine, whether Horizontal, Declining, Including, or Reclining, with severall ways for inscribing the Hower-Lines with ye substile and Meridian.

Navigation. According to ye Plaine Sea Chard and according to Mercator, and by a Great Circle. Also an Exact way of keeping a Reconning at Sea in any kind of Sayling, likewise several ways of finding of a Course, the finding ye Variation [from true North] of ye Compass: a New way of ye professors [i.e. Wakerly's] Invention for ye Accounting of ye variation after a more easy and compendious manner than hath formerly been Taught.

Astronomie. The Principles thereof of Geometrical Demonstracion, Instrumentall Practice and Arithmetical Calculacion, for ye proper Motion of ye Sun, Moone and other Planets with ye Finding of Eclipses for any time past or to com.

Astrologie. The Calculation of Nativities: with ye whole Art of Directions and Annual Revolutions.

The reformation of mathematics

The Tudor revolution in religion was part of a general movement in Europe away from the 'Universal Church' towards more national and secular forms of religion. These were compatible with a new ethic, oriented towards business values. It was now believed that the highest social good was to be achieved by means of a free market, where money could be lent at interest. Previously this had been proscribed as usury, a mortal sin if borrower and lender were both Roman Catholics. Sanctity, once defined as having the right relationship (an unconditional, loving and caring relationship) with God and one's neighbour, was broadened to include devotion to business and profit, and narrowed to exclude people of the lower class, of Catholics and Jews. Success in business was taken to be a mark of divine favour. 'Holy poverty' was recognised, correctly or otherwise, as an inability to come to terms with the divine plan and to cooperate with it. Ordinary poverty was considered to be the result of thriftlessness, a vice that would only be encouraged by the giving of charity.

The Church had long ago established an educational system but this was concentrated, at the secondary level at any rate, on Latin, Christian theology and church music. It paid scant attention to such subjects as mathematics and astronomy. These were 'practical' subjects, needed chiefly to fix the dates of Easter and other movable feasts.

The Reformation saw the first feeble steps towards the process of liberating the intellect, as well as social and economic behaviour, from authoritarian Church dogmas. It opened the way for free inquiry into the phenomena of Nature and for participatory democracy – both congenial for the development of science and mathematics. The process was slow, with many deviations and much backtracking, but by the end of the 17th century a complete revolution had taken place in the physical and natural sciences.

This revolution is generally associated, in England at least, with the name of Isaac Newton (1642–1727) – and the identification is not so far off the mark as such credits are apt to be. Contrary to the mythical figure of Newton, as the simon-pure scientist who became so lost in contemplation of mathematical truths on being awakened that he was discovered hours later still sitting half-clothed on his bed, he was a practical man, involved in finding solutions to the physical and technical problems of the day. For example, he experimented with the transmutation of metals – not, like alchemists before him, to turn lead into gold, but to try to find ways of turning iron into copper, which was at the time badly needed for making cannon, and in short supply. His concern for the defence of Reformed religion led him to stand for election as a member of Parliament for Cambridge. As Secretary and later President of the Royal Society, he encouraged research in scientific matters of every kind. As Master of the Royal Mint, he spent much time and ingenuity combating counterfeiting.

Newton's research in physics systematically covered the full range of topics of interest of the day, making the problems amenable (often for the first time) to the constraints of quantitative methods. Though we concentrate today on his mathematical work, and though it is of almost unparalleled brilliance, it is no more than the tip of the iceberg in a life of outstandingly diverse and extraordinary activity.

The Newtonian calculus

There was a Greek philosophical school at the time of Plato which specialised in inventing arguments that were both paradoxical and disturbing. The school was centred in Elea; its best-known members were Parmenides and Zeno. They were interested in demonstrating that reason demands that we give up some of our most cherished beliefs, many of them so commonplace that they had become part of the language. For example, the idea that many things exist is a fallacy, according to Parmenides. There is only the One. If this is so, then, logically, there are no 'places' occupied by 'things'. If no things exist, there is no need for places to keep them in. The belief that things can move from one place to another is also an error. This, at least, was Parmenides' view.

According to the ancient sources, there were 40 arguments (but if Parmenides is right, why not only one?) which prove that number is an illusion. Two of these 40 original arguments have come down to us, probably in Zeno's own words. They hinge on the statements (i) that whatever has parts cannot be one; (ii) that if each member of an infinitely numerous set has a size greater than zero, then the size of the total set must be infinite.

These arguments are unconvincing. The first is a play on words; the second is irrelevant. However, Zeno's proofs that a moving object is at rest, or that a fast runner can never overtake a slow one (Achilles and the tortoise) are more plausible. The best of them is the paradox of the arrow. 'An arrow is shot into the air. It reaches a certain point (let us call it A). Let us think about it at A. First, it can only move if it is in a place: an arrow cannot move in a place where it is not. Second, it cannot be moving in the place where it is, or it would not be there. Therefore, it cannot be moving.'

This paradox was debated for centuries by the scholastics. The subject of motion (like many other notions about phenomena) was analysed in philosophical terms and in a quasi-religious context. (According to Aquinas, following Aristotle, God was the First Mover, in the sense of setting the universe in motion).

The reasoning of Zeno and the scholastics was based on words. Their activities consisted of constantly defining and redefining what were in fact subjective, descriptive terms. The new interest in science, following Francis Bacon, was concerned with facts. It was inevitable that science would make a clean sweep of the scholastic arguments. In the case of Zeno's arrow-paradox, the calculus (as developed originally by Newton) delivered the *coup de grâce*.

To understand how this happened, we need to be able to talk about the arrow situation in an unambiguous way, and especially not allow ourselves to become 'hung up' on its position as Zeno was. Our first need is for new words, technical terms. A play on words becomes obvious when we have to define (and thus delimit) the meanings of words in preparation for using them. Unlike Zeno, who narrowed the conditions of his argument until he could prove exactly what he wanted, we are interested in everything to do with the history of the arrow's movement, before and especially after it reached its present position.

A few definitions to start with. To represent the distance travelled by the arrow, we normally use the letter s. To represent the time for which it travels we use the letter t. Its velocity (speed), which may not be constant but may vary from time to time, is equal to the distance travelled divided by the time taken: $v = \frac{s}{t}$. If the velocity changes, as it will with an arrow shot into the air, we would use u for the initial velocity and v for the final velocity.

If the arrow is shot straight up in the air, the velocity will change in a systematic way, decreasing to zero as the arrow goes straight up, stopping for an instant as it prepares to change direction, then, as it falls, constantly increasing from zero to some maximum value. The rate at which the velocity changes will be the same when it is decreasing (when the arrow is going up) and increasing (when it is going down). If we ignore slight disturbances in the rate of change due to wind and air resistance, and accept that gravity is the agent that acts (going up) to decrease the speed of the arrow until it reaches zero and then (going down) to increase the speed, then we are talking about the acceleration of bodies falling freely under the influence of the force of gravity. This acceleration has a constant value: an increase or decrease of velocity of roughly 32 feet per second, every second. (It varies from place to place, as the Earth is not a perfect sphere.) It is usually written either as α for acceleration, or as g for gravity.

From the analysis of the motion of bodies moving horizontally, or bodies travelling freely up, or down, under the influence of gravity, and using the symbols defined as above, we can produce three 'equations of motion':

(i) $\quad v = u + \alpha.t$

(ii) $\quad s = ut + \frac{1}{2}\alpha.t^2$

(iii) $\quad v^2 = u^2 + 2\alpha.s.$

(In these equations, the dot means 'multiply'.)

The word 'calculus' has a general meaning as a set of special words, procedures and rules designed to deal with, and to talk about, a special set of problems by means of tested and unvarying rules. The calculus of meanings we use in our ordinary, common-sense conversation is not appropriate in the artificial situation contrived by Zeno (who stops the arrow arbitrarily, or perhaps conceptually is better, in full flight). We need new rules, now definitions and new conventions – in fact, a new calculus such as Newton invented.

A second (invented) example, that of the travelling yo-yo, illustrates the power of Newton's calculus. A certain yo-yo moves in a straight line for some of the time its owner plays with it. During some part of this time it obeys the rule:

$$s = t^3 - 4t^2 - 3t$$

When the yo-yo changes direction so as to come under the control of this equation, for an instant its velocity is zero. The question is, what is its acceleration at this point?

We are talking here about instantaneous velocity and instantaneous acceleration. In other words, the yo-yo is actually at rest (its velocity is zero), but this is not its final state, since its acceleration at this point is not zero. Like Zeno's arrow at its highest point, it still has energy stored up, so that at the appropriate moment it will resume its motion. It has reached a limit, but this is only a temporary state. The value of the instantaneous velocity is obtained by means of Newton's differential calculus. We differentiate distance by time: that is, we determine the 'instantaneous' rate of change of the position. Since velocity $= \frac{s}{t}$, the instantaneous velocity $= \frac{ds}{dt}$ (that is the derivative of distance by time). (We are now using Leibniz's notation. Newton's notation for the differential calculus is obsolete, except in some parts of the USA.) This expression $\frac{ds}{dt}$ means that the velocity at this instant is the very short distance travelled by the yo-yo in the very short instant of time (so short that you can call it infinitely short – it is almost zero, but not quite.)

At this point, we move into Newton's differential calculus. This tells us that:

$$\text{If } v = \frac{s}{t} = \frac{(t^3 - 4t^2 - 3t)}{t}$$

then, differentiating,

$$V = \frac{ds}{dt} = 3t^2 - 8t - 3$$
$$= (3t + 1)(t - 3)$$

This means that the instantaneous velocity is equal to zero when $(3t+1)(t-3)=0$, that is, when $t=3$ or $t=-\frac{1}{3}$.

By the same line of argument, the instantaneous acceleration is

$$\frac{dv}{dt}.\ \left(a=\frac{v}{t};\ \frac{da}{dt}=\frac{d^2s}{dt^2}\right)$$

In other words, we differentiate the velocity by time and differentiate the distance by time twice to obtain instantaneous acceleration.

In other words, for the yo-yo, $a=\frac{dv}{dt}=6t-8$. We substitute 3 for t and then $-\frac{1}{3}$ for t in this equation, since at 3 seconds and $-\frac{1}{3}$ seconds (as shown above), the velocity is zero. Then $a=10$ or -10 when $t=3$ or $-\frac{1}{3}$, respectively. (We can ignore the negative value except perhaps to make the mental note that, like Zeno we have not yet got the full story. If we are interested in a complete statement about the yo-yo, we must record that the velocity is zero also at $\frac{1}{3}$ of a second before our original observation, the acceleration then being 10 feet per second – but, as it is negative, it is actually a deceleration and the yo-yo is slowing down. Where Zeno went wrong is that there was no way open to him to recognise what we have called the 'instantaneous velocity', meaning the actual velocity of a moving body measured in an infinitely small unit of time when it is momentarily at rest.)

The binomial theorem

The binomial theorem, discovered by Newton when he was a student in his early twenties, allows us to write down the expansion of $(x+y)^n$, where n is a whole number and x and y are two unknown numbers. We can write down, in order, all the terms in powers of x and y.

The need to use this expansion might arise if, for example, you were gambling, for instance playing what is called 'heads and tails' or 'pitch and toss'. The object of this game, in its simplest form, is to guess how many coins will land heads uppermost when 10 coins are tossed. Obviously, a player who knows beforehand what the most likely combinations are and how many times they are to be expected will have an advantage.

We can represent this situation by considering the expression $H+T$ (where H represents the number of heads and T the number of tails). Clearly, if one coin is thrown, there are only two possibilities: one H or one T. For two coins, the possibilities are HH, HT, TH and TT; there is only one way to get either both heads (HH) or both tails (TT), but there are two ways to get one head and one tail, that is, HT and TH. If we look at the expansion of $(H+T)^n$, where n is the number of coins, we see the same situation. $(H+T)^2=H^2+2HT+T^2$, and so we get the same result as

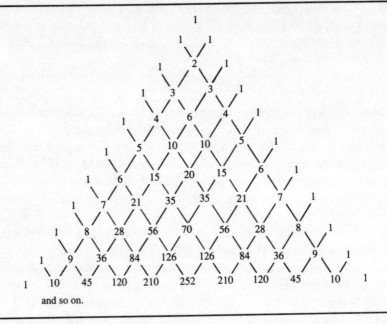

and so on.

previously. As we increase the number of coins, it becomes tedious to write down every possibility and count the number of heads in each. (With 10 coins, for example, there are 1024 possibilities.)

This is where the binomial theorem comes in. It tells us how to find the 'coefficients' (the number of each term in the expansion). These numbers are given by 'Pascal's triangle'. This starts at the top with a triangle made up of three 1s, to which new rows are added by writing 1s at each end, and, in-between, writing the sum of the two numbers immediately above.

These give the coefficients very simply: starting from the left of the 10th row, we have, in order, the coefficients of $H^{10}, H^9T, H^8T^2, H^7T^3$ and so on, down to HT^9 and then T^{10}. Counting along, we see that the coefficient of H^5T^5, 252, is largest; so our gambler is most likely to be right 262 times out of 1000 if he or she guesses 5 heads every time. Thus, the accurate knowledge of possibilities is an advantage to the gambler.

Pascal's triangle is a mechanical routine which gives the answer. Newton discovered a formula which provides the answer in a much more elegant way. The binomial theorem is used in the theory of permutations and combinations, and so also in the theory of probability. The probability question we are answering is, In how many ways are you likely to get heads (or tails) if you toss 10 coins in the air 1024 times, that is 2^{10} times? What Newton found was a method for working out these coefficients, without having to draw out a large table and perform many operations to discover what they are. It is easiest to see by means of an example, so we will again look at the coefficients of $(H+T)^{10}$.

The coefficient of H^{10} is given by the following:

$$\frac{10 \times 9 \times 8 \times 7 \times 6 \times 5 \times 4 \times 3 \times 2 \times 1}{1 \times 2 \times 3 \times 4 \times 5 \times 6 \times 7 \times 8 \times 9 \times 10} = 1 \text{ (by cancelling)}$$

To find the next coefficient, we write the same fraction down, but leave out the two numbers at the right-hand end. Thus, the coefficient of $H^9 T$ is:

$$\frac{10 \times 9 \times 8 \times 7 \times 6 \times 5 \times 4 \times 3 \times 2}{1 \times 2 \times 3 \times 4 \times 5 \times 6 \times 7 \times 8 \times 9} = 10$$

We do the same again to find the coefficient of $H^8 T^2$:

$$\frac{10 \times 9 \times 8 \times 7 \times 6 \times 5 \times 4 \times 3}{1 \times 2 \times 3 \times 4 \times 5 \times 6 \times 7 \times 8} = 45$$

and so on.

We can, in fact, straightway write down the coefficient that we want. Suppose, for example, we want to find the coefficient of $H^6 T^4$. On the bottom (denominator) of the fraction, we write the product of all the whole numbers up to 6 (this is called 6 factorial):

$$1 \times 2 \times 3 \times 4 \times 5 \times 6$$

Then on top (the numerator) we begin at 10 and work downwards, until we have written down the same number of numbers as we have in the denominator:

$$\frac{10 \times 9 \times 8 \times 7 \times 6 \times 5}{1 \times 2 \times 3 \times 4 \times 5 \times 6}$$

Then we cancel until we find the number we want:

$$\frac{10 \times \overset{3}{\cancel{9}} \times \cancel{8} \times 7 \times \cancel{6} \times \cancel{5}}{1 \times \cancel{2} \times \cancel{3} \times \cancel{4} \times \cancel{5} \times \cancel{6}} = \frac{10 \times 3 \times 7}{1} = 210$$

This works for any power of $(H+T)$; for $(H+T)^n$ we start the numerator at n. So, for example, in the expansion of $(H+T)^{200}$, the coefficient of H^7T^{193} is:

$$\frac{200 \times 199 \times 198 \times 197 \times 196 \times 195 \times 194}{1 \times 2 \times 3 \times 4 \times 5 \times 6 \times 7}$$

$$= \frac{\overset{100}{\cancel{200}} \times 199 \times \overset{33}{\cancel{198}} \times 197 \times \overset{\overset{7}{\cancel{28}}}{\cancel{196}} \times \overset{\overset{13}{\cancel{39}}}{\cancel{195}} \times 194}{1 \times \cancel{2} \times \cancel{3} \times \cancel{4} \times \cancel{5} \times \cancel{6} \times \cancel{7}}$$

$$= \frac{100 \times 199 \times 33 \times 197 \times 7 \times 13 \times 194}{1}$$

$$= 2,283,896,214,600$$

Newton's discoveries in mathematical physics

Newton's classic text was *Philosophiae Naturalis Principia Mathematica* ('The Principles of Mathematical Philosophy'), written in Latin and published in 1687. In the first chapter, he declared what are now known as Newton's Laws of Motion. These state that:

(i) Every body continues in its state of rest or of uniform motion in a straight line unless compelled to change that state by some external, imposed force.

(ii) The force acting on a body in motion is directly proportional to the mass of the body multiplied by its acceleration. (This is normally put in the form $F = ma$.)

(iii) To every action there is an equal and opposite reaction.

Newton declared these not as 'laws' but as 'axioms'. He believed them to be not empirical laws, such as we could demonstrate in a laboratory, but more like definitions of what is meant by 'force' (that which causes change of motion in a body), 'mass' (the amount of matter in a body) and 'acceleration' (that rate of change of velocity). However, we now have simple pieces of apparatus that can demonstrate Law (ii); the other two are more like definitions.

In the same text, Newton set out his other great discovery, the universal law of gravitation. This states that 'Every particle of matter in the universe attracts every other particle with a force F which varies directly as the product of their masses (m_1, m_2 respectively) and inversely as the square of the distance d between them.' This is written as

$$F = \frac{Gm_1 m_2}{d^2}$$

where G is a constant, varying from place to place, but known as the 'Universal Constant of Gravitation'.

Using these four principles, Newton developed a model of the universe which explained how it worked under the mechanical control of the forces acting on it. Thus, the laws applied universally, in that they accounted for physical phenomena not only on Earth, but in empty space and all other objects everywhere.

Unlike Laplace (who told Napoleon that he had no need for the 'God hypothesis' to explain celestial mechanics), Newton believed that God was essential to explain the design, creation and maintenance of the system as a going concern. It was the initial impulse provided by God (the First Mover), working in conjunction with the force of gravity (invented by God), which maintained the planets in their orbits. Two centuries later, this doctrine, of God as the First or Prime Mover, was succeeded by the Kant–Laplace hypothesis. This took further Newton's argument that physical forces alone were the source of planetary motion. In the case of the solar system, the initial impulse was identified by the philosopher Kant (and the explanation was taken further by Laplace) as the outcome of a rotating and cooling nebula. As it cooled it contracted, and as it contracted it spun faster and faster due to the law of conservation of momentum. It reached the point where great masses of molten rock were thrown off due to increasing centrifugal force. These masses solidified to become the planets which continue to circle around the Sun.

The Committee on the Longitude

In 1707, an English fleet was wrecked on the Scilly Isles in the English Channel. This was an especially dangerous area and, on this occasion, the Admiral, Sir Cloudsley Shovell, and a large number of sailors were drowned. The tragedy led to several years of public debate. The problem was the lack of an accurate way of calculating longitude – something that also severely hampered long-distance navigation. In 1714 the British government offered a reward of £20,000 (several million, in today's values) to anyone who could come up with a satisfactory solution to this problem; the French government offered a similar sum in the following year.

This problem stirred the scientific community. Within a year, 25 proposals were submitted, and the Committee on the Longitude was set up to assess them. It was made up of scientists – Newton (President of the

Royal Society), Sir Edmund Halley (Secretary of the Royal Society), John Flamsteed (Astronomer Royal), Keill, Sanderson and Cotes (Savillian, Lucasian and Plumian Professors of Mathematics at Cambridge) – seamen (the Lord High Admiral, the First Sea Lord, the Admirals of the Red, White and Blue Squadrons) and one politician (the Speaker of the House of Commons).

Since the Earth is a sphere, when you travel all round and return to your starting point, you pass through 24 one-hour time zones. So 360 degrees of longitude is equivalent to 24 hours of time, and vice-versa. Thus, if we know the local times at two points on the Earth's surface we can easily fix their relative longitudes.

The chief problem was accurate measurement of time. The technology of time-keeping was still, in the 1710s, in a rudimentary state. Pendulum clocks were a new invention, still being worked on by scientists such as Huygens and Robert Hooke. Watches, driven by a main-spring, were known but were not common. Travellers' time-pieces consisted of a gnomon, or tiny sun-dial, set up in a pocket-watch which could be opened and set with the help of a compass. Hour-glasses were part of most ships' normal equipment.

When the shouting had died down and the money had been handed over, it was generally accepted by those most competent to judge that the ideal solution of the longitude problem was a precise clock which could be compensated so that it could be used in tropical and cold climates.

The ideas put forward, in all seriousness, to the committee, show that scientific method at the time was almost as little understood (or at least applied) as it had been in the days of the ancient Greeks. Some of the choicer proposals were:

(i) Isaac Hawkins: a Barometric Method

A barometer would be used, in conjunction with a globe, to draw up a set of universal tide tables showing high and low tides. The reading of the barometer at a certain position at sea would yield the levels of high and low tides at that place. The longitude could then be found by consulting the tables. (The method remains obscure and seems to suggest that Hawkins had probably never actually used, or seen, a barometer).

(ii) E. Place: time signals

It was proposed to build a powerful lighthouse which would flash time signals on the clouds. Ships up to 200 miles away would see these and could compare their (local) time with the time signalled at each lighthouse. The latitude and longitude of the lighthouse having previously been determined with great accuracy, the ship's longitude could then easily be obtained.

(iii) *Measure the time accurately*

(a) by a watch driven by a mainspring (Sebastiano Ricci).
(b) by a watch heated from beneath (Stephen Plank). This solution addresses the nub of the matter: the need for an accurate clock or watch that compensated for tropical climates and for freezing temperatures. Unfortunately, its proponents knew of no such timepiece, and had no suggestions about how to set about making one. It was not until 1758, after 33 years' work, that John Harrison invented exactly the kind of marine chronometer they suggested, but without the fire down below.

(iv) *The winners – William Whiston and Humphrey Ditton: sound and light signals*

This method consisted of having a network of hulks anchored at sea in known latitudes and longitudes. These would fire cannon, presumably according to an advertised schedule. Ships at sea in the immediate neighbourhood would first detect the flash of the explosion and then, some seconds later, would hear the noise of the explosion. The velocity of sound being known (approximately 1078 feet per second), the distance could be calculated by multiplying the time between the flash and the noise of the explosion, by this constant. If three ships were suitably placed, their observation of the shots would give a triangulation from which an accurate position could be calculated.

Two questions immediately occur to the practical seafarer. First, what happens when there is a storm blowing, or even just some very strong winds? The velocity of sound in air would certainly be affected by winds and turbulence in the air which carries the sound. Second, this method might be fine for calculating the longitude for places anywhere within about 85 miles of the stationary hulks, but what about the rest of the oceans? In spite of such objections, Whiston and Ditton were awarded the prize money, with only Newton expressing his reluctance.

The whole affair – especially viewed in the light of modern navigational methods – may seem ridiculous. But it shows two interesting facts about the 'climate' of science at the time, both of them marking advances on what had existed only a century before. First, the large response to the competition indicated that a solid infrastructure for scientific development existed in Britain. There were considerable numbers of people actively involved in teaching navigation, in making scientific instruments, and in providing theoretical and practical training for careers in mathematical and scientific work.

Second – an even more remarkable development – theoretical scientists of the first rank, such as Newton, Flamsteed, Hooke, Halley and Briggs, were prepared to work in a team enterprise with shopkeepers, manual workers, tradespeople and ordinary seamen. The team spirit manifested was able, to a degree unknown before the 'scientific revolution' of the 17th century (of which Newton was a prime instigator), to overcome the normal English consciousness of class and religious differences and the difficulties of cooperation between classes.

BABBAGE, THE GREAT UNKNOWN

Hats off, gentlemen: a genius.
Schumann to students, on hearing Chopin play

Man wrongs, but time avenges.
Byron

CHARLES Babbage was born in Devon in 1792; he died in London in 1871. He went up to Trinity College, Cambridge, to study mathematics and soon discovered that he knew more about it than his tutor. At 20, with John Herschel (later to rival his father as an astronomer) and George Peacock (later Dean of Ely) he set up the Analytical Society to 'awaken the Mathematicians from their dogmatic slumbers'. The initial objective was to abolish Newton's calculus notation in favour of Leibniz's. Indeed, the three undergraduates wrote a book devoted to proving that Leibniz's 'd' notation ($\frac{dy}{dx}$, standing for the derivative of x) was better than Newton's 'dot' notation (that is, $\dot{x}$ which stands for the same thing).

At 24, Babbage was elected to the Royal Society. (Since his day, this has become the highest honour open to scientists in Britain. Babbage must be given most of the credit for this change. In his time, it might merely indicate some aristocratic connections.) When he was 25, he founded the Royal Astronomical Society. At 36 he was elected to Newton's chair in Cambridge, and was Professor for 11 years before he resigned. He never lived in Cambridge and never gave a lecture. In fact, he accepted the job only to please those friends who had nominated him. His main duties were to examine candidates for honours and to award prizes in mathematics – and he used both to encourage higher standards in the profession.

On one occasion, while Babbage and Herschel were still students, they were working together to correct tables of logarithms. These tables had long been notorious for mistakes – and recalculating and correcting them was soul-destroying work. The task demands so much accuracy, and at the same time is so mindless and time-consuming, that it is more suited to machines than to human beings. Babbage remarked to Herschel, 'It's a pity this can't be done by steam' – an idle comment, but one that seized hold of him to such a degree that it dominated the rest of his life.

Babbage began to harness machine power to calculation. In those days, before the common use of electricity to drive machinery, this meant building a spring-driven or weight-driven device with thousands of intricate moving parts. The precision engineering that was required was costly. Babbage began by building a small 'difference engine' (a mechanical calculator) and then, rashly, applied for a government grant to build a larger one. He soon found that this grant was not enough. Having spent a large part of his own private fortune, he applied to the government for more funds. During the waiting time he designed an even more elaborate machine, a forerunner of the modern computer, which he called the 'analytical engine'. It made the concept of his difference engine obsolete. He wrote to advise the government of this development.

Babbage wanted the authorities to say whether he should finish the deluxe model of the difference engine (for which he had received the earlier grant), or forget about it. But the bureaucrats seemed to become confused between the two machines. They seemed incapable of working out, or remembering, which machine he was engaged in building. He failed to make them understand that he had already built the difference engine which was completely operational. (It survives, in working order, in the London Science Museum.) They came to no decision, and kept him dangling for years. Eventually, in 1843, many years after Babbage's first application, the money was refused by Disraeli, then Chancellor of the Exchequer.

Charles Dickens, a friend of Babbage, devoted a large portion of his novel *Bleak House* to one government department, the Court of Chancery, which he named 'The Circumlocution Office'. This court dealt with the probate of contested wills. Often, those contesting a will, and the legal heirs, had to wait anything up to 40 years for a decision to be handed down. By this time, two or three generations of lawyers' fees had eaten up the inheritance; most of the litigants were dead; and the deceased's intentions had been totally subverted. It is more than likely that Dickens had Babbage's case in mind when he wrote *Bleak House*.

Babbage's interests were by no means confined to mathematics or computing machines. He was a man of universal curiosity, a problem-solver. He sponsored the penny post in Britain, analysing deliveries of the contents of Bristol mailbags for a week and demonstrating that it would be more efficient and economical if all letters were to go at the same rate (one 'old' penny), regardless of distance. He invented the cowcatcher to solve the problem of cattle, and people, straying in front of moving trains. He invented railway signals and the speedometer. He devised a system of signalling by lighthouses (first used in the Crimea by the Russians against the British, who had shown no interest). He claimed that there was no lock that he could not pick, no code that he could not decipher. He invented games theory to teach his analytical machine to play chess. He drew up plans, and spent a fortune on blueprints, for gears and other machine parts of a microscopic precision unimaginable at the time – and trained a group of workers to produce them. He made a survey of all engineering works and manufacturing methods in Britain, and wrote an influential book about it.

Babbage gave a series of lectures in Turin outlining his ideas for the new kind of calculating machine. These were listened to avidly by the Italian scientists. An army man, Major Menabrea, was particularly impressed, so Babbage supplied him with copies of his published works, a set of blueprints for the machine and some private papers. Using this material, Menabrea wrote an account in French for a Swiss journal. This was translated by Ada Augusta Lovelace (Lord Byron's daughter), and published in England. Ada Augusta's editorial notes were more than twice as long as the original paper. This document is the main source for Babbage's ideas about the analytical engine at this time. (Ada Augusta is often credited with writing the first published computer program. With Babbage's help, she wrote a complex program for machine calculation of Bernoulli numbers (an infinite series important in the theory of probability), and published it in her 'notes' to Menabrea's paper. Thus, she shares the priority with Babbage, who invented the idea of programming in the first place: see below, page 196.)

Babbage had two idiosyncrasies, one being that he told the truth regardless of the personalities involved. These included most of the grandees of the English science community. For example, when he was upset by the conduct of the Astronomer Royal, and by the President of the Royal Society, Sir Humphrey Davy, the great chemist of the age, he said so, in detail and publicly. He wrote a book about how the Royal Society was destroying science in Britain: a scandalous charge, not least because it was true. Babbage's other major character trait – or perhaps one should say eccentricity – was a pathological hatred of organ-grinders and other street 'musicians'. He campaigned against them, claiming (as a mathematician whose main interest was in industrial efficiency) that they cost him 25 per cent of his working time. They responded to his irate comments, made face to face, by deliberately serenading him, following him about and performing under his window wherever he happened to be.

The difference engine

There is a rule in science, not quite a law, which states that if a number of people work on a problem, credit for any discovery will almost invariably be given to the best-known of them. This is true, for example, of 'Pascal's triangle'. It is also true of the binary system. This was invented by the Chinese and taken over (via 17th-century Jesuit missionaries to China) by Leibniz. At least a dozen Europeans had advocated its adoption before his time. This rule about priority (which John von Neumann described as the ability of notable persons, like himself, to emerge first from a revolving door, having gone in last), does not apply to Babbage's engines. The idea that machinery could be a realistic basis for mass production of solutions to number problems was unique to Babbage.

Yet, although his idea was original, it owed its inspiration to the French. When the metric system was adopted during the French Revolution, it presented the task of recalculating dozens of tables to replace the now obsolete system of weights and measures. Two prominent French scientists, Proly and Legendre, were put in charge of the work. They organised the task so that six expert mathematicians (themselves included) worked out the formulae needed. Another group of eight less important scientists then worked out the key numbers and monitored execution of the plan. Still a third group, 60 to 80 strong, did the donkey work of extrapolation (or rather interpolation), mechanically filling in gaps in the tables.

It was the factory mass-production method applied to science – and hence appealed particularly to Babbage, with his interest in streamlining industrial processes. His interest focused on the question whether the 60 to 80 'mechanics' could be replaced by a machine. The French mathematical helots needed no other knowledge than how to add, subtract and divide by two. Babbage's intended machine would be able to perform similar simple tasks. But it also would be part of a grand design which would allow more complex calculations. The machine would never tire, would make no mistakes except when some part wore out or the human operator gave it a wrong number or instruction.

In creating such a machine, the first requirement is an algorithm to spell out the problem step by step. This must lead by steady progress towards the final solution. A simple example of how such an algorithm works is seen in calculating the terms of a number series: those, say, that continue the sequence 1, 4, 9, 16.

A human working on this problem might realise that each number in the series is the square of a number in the ordinary decimal sequence: 1 is (1×1), 4 is (2×2), 9 is (3×3), 16 is (4×4). The series can be continued by the same process: $(5 \times 5) = 25$, $(6 \times 6) = 36$ and so on. To realise this is the result of 10 years of instruction in mathematics. The question for Babbage was: How do you set up a machine to do this and similar tasks?

A method commonly used in dealing with number series of all kinds involves writing the original series down horizontally, then writing it again on the next line, moving each number one space to the right, then subtracting the new sequence from the original series. This produces 'differences'. The process is then repeated, again and again if necessary, until the differences (the results of subtraction) are all zero. In the series 1, 4, 9, 16, the method works as follows:

Original series	1	4	9	16
Series moved one space right ('first degree')		1	4	9
'Difference': result of subtraction		3	5	7
Moved one space right ('second degree')			3	5
'Difference': result of subtraction			2	2
Moved one space right ('third degree')				2
'Difference': result of subtraction				0

Note: This applies to the series however far you prolong it. The bottom line is always zero.

The table above can be written more simply as follows:

Original series	1 4 9 16
Differences	3 5 7
Second differences	2 2
Final difference	0

Babbage knew that if you produce difference series in this way, the number of times you have to subtract to arrive at zero differences tells you the rank or power of the original series. In this case, the original series (1, 4, 9, 16) is made up of numbers of the second degree.

The general form of this expression is written as

$$\text{Number} = a.x^2 + b.x + c$$

where a, b, and c are constants, and x varies to take the values 1, 2, 3, 4, . . . in succession. (The dots in the equation are multiplication signs.)

Term	2nd Diff.	1st Diff.	Term value	Remarks
1st			1	starting point
2nd	2	1	4	$(2+1+1=4)$
3rd	2	3	9	$(2+3+4=9)$
4th	2	5	16	$(2+5+9=16)$

Now continue:

Term	2nd Diff.	1st Diff.	Term value	Remarks
5th	2	7	25	$(2+7+16=25)$
6th	2	9	36	$(2+9+25=36)$

etc.

At this point, Babbage noticed that we can reverse the procedures and so reconstruct our original series. To do this, it is enough to know the two difference series. Then – and this is the point of the operation – by using this algorithm we can extend the original series much further. We know our starting point (in this case, the number 1); we know that our first difference series starts at 1, then goes to 3, then to 5, then 7, and so on; we know our second difference series (which is simply 2, 2, 2, repeated as necessary). If we now add these series, we can continue writing the original series as far as we wish. The results and method are shown in the table above.

Before describing the machine, we will use the algorithm we have discovered to write instructions for it. Since the instructions are in English, they are intended for the operator. ('Reminder' is a comment or command for the operator; it is not needed, nor can it be used, by the machine.) The instructions start by assuming, once again, that we know nothing except the first four numbers; 1, 4, 9, 16.

All commands were given to the machine by the operator by means of cards threaded together in correct sequence. This system was originally devised by the French entrepreneur Jacquard, to program mechanical looms which wove coloured patterns in textiles. The pattern was given to the machine by means of punched holes in cards which were so placed as to allow long needles (from a 'bank' arranged like bristles in a stiff brush but able to move in and out) to pass through the corresponding holes. Where there were no holes, the needles were prevented from passing through. The pattern of holes on the cards, and therefore the pattern of needles emerging from the holes, operated shuttles on the loom to bring different-coloured threads into play and so change the pattern woven into the cloth, step by step. In Babbage's engine, in place of shuttles, the commands activated gears and wheels (with engraved numbers for the benefit of the operator).

Reminder: First step, find the power of the series. To do this, write down the various difference series until all the numbers are the same.

01. Set up first four numbers   *Reminder*: 1, 4, 9, 16
02. Subtract each from next   *Reminder*: 1, 3, 5, 7
03. Repeat for 2nd diff. ser.   *Reminder*:   2, 2, 2
Reminder: Series is of the second order (squares)

1st difference series is 1, 3, 5, 9
2nd difference series is 2, 2, 2, 2 . . .
Starting number of original series is 1
Continue

Reminder: Proceed to the second step, set up the machine to calculate (say) 1000 terms of the series.

04. Print known numbers of original series (1, 4, 9, 16).
05. Remove all numbers (i.e. set all registers to zero).
06. Set up new numbers:
      16 (next term); +7 (1st difference); +2 (2nd difference).

Stand aside until machine has finished.

Reminder: The machine has been preset by operator to do a recursion operation (see below). The addition card will instruct it to add where necessary to produce the next numbers, beginning with $25+9+2=36$.

07. Transfer these numbers 25, 9, 2 to "memory" (store).
08. Re-enter next numbers; 36, 11, 2; starts new cycle.
09. Go back to step six; repeat steps 6 to 9; 995 times.

Reminder: The last instruction starts a loop for recursive processing (repeating the operation as often as required).

The punched cards were in accord with the algorithms for the required calculation. The operator's tasks were to select the necessary cards from a card-file, string them together in correct sequence, attach them to the machine, and set up the initial state of the machine and the numbers required to initiate the calculation.

The analytical engine

Babbage's prototype for the difference engine was exhibited, among other places, at the 1851 Exhibition at the Crystal Palace in London. For some reason known only to the Exhibition Committee (probably the antagonism towards Babbage by Airey, the Astronomer Royal and chairman of the Committee), it was tucked away in a basement and almost inaccessible to visitors, as though it were a state secret. It made little impression. But in any case, Babbage had already decided to replace it with the far more intricate 'analytical engine', able to perform any calculation whatsoever, so long as the calculation could be broken down into simple, unambiguous, sequential instructions.

Babbage's analytical engine was not completed in his lifetime, although he spent a fortune on plans and precision gears. It was finished after his death by his son. There is no way of knowing what refinements he might have introduced at the construction stage.

In the case of the illustration of the series of square numbers described earlier, the engine would obtain and print out the sums of the difference series. This means adding two difference series to each new number as it emerges, to obtain the next number. (This is the process of recursion, the generation of new numbers in a continuing process, in which the same mechanical routine is applied again and again, as often as necessary.) The chart below allows us to follow the numbers as they go through the machine and appear on the dials. (In the second line of the chart, d.s. stands for 'difference series'.)

Starting position:		1	1	→ Prints 1
d.s.1	d.s.2	Series 1	Series 2 (given)	
2 (+)	1 (+)	1 =	4	→ Prints 4
2 (+)	3 (+)	4 =	9	→ Prints 9
2 (+)	5 (+)	9 =	16	→ Prints 16

Suppose we start in the middle, with the term 9. We want to see how the machine gets the next number in the series.

Dial C is set to 2	Dial B is set to 5	Dial A is set to 9	Dial D is set at 0
Hammer C	Hammer B	Hammer A	Printer D

Numbers are transferred one at a time from one dial to the next, left to right. Thus, Dial C reads 2 at the start; its hammer strikes twice. At each strike dial C moves on one number: from 2 to 1, and from 1 to 0. As dial C units disappear, they are transferred to the next dial. They appear on Dial B, which moves to 6, then to 7 as the numbers are transferred.

1. strikes 1 → now reads 6   —   —
2. strikes 1 → now reads 7   —   —
Dial C is now at zero
Dial B now stands at 7
Dial A has not changed
Hammer B now strikes 7 times
Dial A now moves along by ones to add 7

1. strike 1   → reads 10   —
2. strike 1   → reads 11   —
3. strike 1   → reads 12   —
4. strike 1   → reads 13   —
5. strike 1   → reads 14   —
6. strike 1   → reads 15   —
7. strike 1   → reads 16   —

Hammer B is now at zero. It stops.
The number 16 then passes from A to D.
Dial D now gives the total to the printer, which prints 16.

The numbers are also recorded in a duplicate set which records these changes to form control figures, which are reset by the machine from the memory, on dials A, B, C, D. This being done, the process continues, generating the next number. In this case, the number 2 is passed from dial A to dial B to dial C to dial D to the printer.

In this phase dials are set at the start at:

$$2 \quad 7 \quad 16 \quad 0$$

The next phase adds on 2 and passes it along, as follows:

$$2 \to 0 \quad 9 \to 0 \quad 25 \to 0 \quad \text{Prints 25}$$

The process continues, to generate the next number.

The invention of programming and machine language

When Jacquard created punch-card 'programs' for his looms, he was in effect converting the loom into a robot, programmed to select different coloured threads which would produce patterns in the cloth being woven. Babbage thought that exactly the same process could be used in solving problems: the machine (in Ada Augusta's metaphor) could 'weave' different patterns of numbers. The machine would go through the different operations required and print out the answer. Unlike Leibniz's calculating machine (which was not fully automatic, every number being cranked out place-by-place by the operator, who had also to reset the machine for each stage of the calculation), the analytical engine was designed to be entirely under the control of the program. Once it was set up and started, it would finish the job. The program (of punched cards wired together) would call on the machine to make decisions, to 'reach in' (if instructed by the program to do so), and change the routines.

The program called for an orderly set of activities, clearly specified. This involved meticulous description of the essential steps for solving the problem, linked in the correct sequence, and described with absolute clarity. The program had to be unambiguous and self-consistent. At each step, the machine had to be able to work out the numbers needed for the next step. In other words, in writing a program, Babbage had to foresee all the possible ways in which the machine could go wrong because the instructions were not clear or could not be carried out. The program had to be able to produce exactly the machine behaviour that he intended. Goethe's story of the Sorcerer's Apprentice gave dreadful warning of how omission of just one essential step might be a recipe for chaos.

To communicate with the machine, Babbage had to invent a new language, a kind of symbolic shorthand based on an analysis of machine movements. The idea of a 'universal language' came from Leibniz. But is was Babbage who applied it to machines, so inventing the whole concept of 'machine language', in which we speak to machines, they speak to one another, and they communicate with us.

The analytical machine and complex problems

It is possible to get some idea of the elegance of Babbage's programming, and of the complexities of which his analytical machine would have been capable, if we see how it would have attacked the problem of solving simultaneous equations. (This is the same problem, with the same amount of sophistication, as was solved by non-mechanical means in the Chinese text of the 6th century BC, described on pages 58–60. Various algorithms or routines can be used to solve it and others like it. The first machine solution – by Ada Augusta and described below – is given in her notes to Menabrea's 1842 account of Babbage's work. It enables us to solve any two simultaneous equations by entering only six numbers into the machine. To explain how this is done, we need to look at the general form of the equation:

$$ax + by = p \qquad \text{(equation 1)}$$
$$mx + ny = q \qquad \text{(equation 2)}$$

The numbers a, b, m and n are the multipliers of x and y and are called coefficients; p and q are the results of multiplying the xs and ys by these numbers, and adding the results. The problem is to find the values of x and y. To solve for x, we need to get rid of y from the equations; similarly, we can find y by getting rid of x. If we multiply equation 1 by n, and equation 2 by b, we then have two new equations in which the coefficients of y are the same (nb).

$$anx + bny = np \qquad \text{(i)}$$
$$bmx + bny = bq \qquad \text{(ii)}$$

Now if we subtract (ii) from (i), the ys disappear:

$$anx - bmx = np - bq$$

that is,

$$(an - bm).x = np - bq$$

Therefore

$$x = \frac{(np - bq)}{(an - bm)}$$

A similar process gives the formula for y:

$$y = \frac{(aq - mp)}{(an - bm)}$$

This way of finding x and y is the kind of operation for which machines are ideal. To find the values of x and y, we need to tell the machine six numbers: a and b; m and n; p and q. We must also provide it with a program that will allow it easily to keep track of manipulations with the input data $(a, b, p; m, n, q)$. We can set it up to solve any valid equations by telling it when (and what) to multiply; when (and what) to subtract; and when (and what) to divide.

The program we are describing works by manipulating the variables x and y. The numbers a, b and m and n are coefficients; p and q are ordinary numbers. Each equation will have different values for a, b, m and n. As these values change, so do the values of x and y. Each of the required operations (addition, subtraction, multiplication and division) is presented to the machine, at the exact point it needs them, by punched cards, as in the Jacquard loom.

For the whole operation Babbage therefore needed only eight cards:

    6 cards for the values of a, b; m, n; p and q;
    1 card giving instructions on how to subtract
       (this card is presented to the machine three times);
    1 card to instruct the machine how to divide
       (this card is used twice).

The first six cards consist of the four coefficients of x and y and the two

numbers p and q. These six pieces of data specify the equation. The other
two cards control three subtraction operations and two divisions.

At first sight, the machine program that follows may seem very long. But
this is because the calculation has been broken down into tiny steps – and it
is no longer than describing how to do even the simplest operation (e.g.
subtraction) using paper and pencil, or how to use the Chinese algorithm
on page 59. The machine, too, unlike a human mathematician, was
designed to carry out these operations tirelessly, 24 hours a day, without
error.

The data numbers, as they go into the machine, have to be assigned a
local habitation (known as an 'address') and a name. (Any name will do,
provided that is is used throughout to identify the same variable.) Without
these two attributes, they might be confused with other variables. In this
program, names were assigned by Babbage. He describes them all as
variables, using the letter V, and gives each V a different number, from V_0 to
V_{13}. To avoid confusion, I have altered his subscripts to read V(1) to V(14)
and added two more, V(15) and V(16) to show the output. But I have
otherwise kept the exact structure of his argument.

This particular algorithm needs six numbers to start it off. These are the
four numbers used as multipliers of x and y (the coefficients) as well as the
two numbers for m and n. Each of these variables is kept separate from the
others, assigned a name, V(1), V(2), and so on, and a place on a particular
wheel in the machine. (Babbage calls the wheels 'discs', but they seem too
large and solid for such a description.)

The variables are 'read' on to the wheels by the human operator: in other
words, he or she sets the wheels by hand to start the process. For example,
our six starting points (or pieces of data) are as follows:

Input: $V(1) = a$ $V(2) = b$ $V(3) = p$
            $V(4) = m$ $V(5) = n$ $V(6) = q$

(a, b, m, n, p and q are, of course, Arabic numbers, not letters, when we are
solving an actual equation.)

The operator arranges the program cards to deal with these data,
combining and recombining them as the algorithm prescribes. The
machine, in fact, will generate the rest of the data needed to solve the
problem. First, it will multiply the variables in pairs, as needed. Then these
products will be subtracted in defined ways. Finally, there will be two
divisions, and the values of x and y, so calculated, will be printed out.

The machine does all these operations in sequence, under the control of
the program. As the wheels revolve, and as the different variables are
brought into existence, the numbers appear as follows:

Multiplying: $V(7) = an$ $\quad V(8) = bm \quad\quad V(9) = np$

$\quad\quad\quad\quad\quad V(10) = aq \quad\quad V(11) = mp$

Subtracting: $V(12) = an - bm \quad\quad V(13) = np - aq$

$\quad\quad\quad\quad\quad V(14) = aq - mp$

Dividing: $\quad V(15) = \dfrac{V(13)}{V(12)} \quad\quad V(16) = \dfrac{V(14}{V(12)}$

Printed answer: $x = V(15) \quad\quad y = V(16)$

Thus, in solving the problem, the machine has to create 10 new variables. Six were provided to start with. This means that 16 of the grooved, metal wheels are engaged in the operation. The actual calculations are carried out in another part of the machine, which Babbage called the 'mill' (in a modern computer it is known as the central processing unit, or the arithmetic unit: see pages 236–8). The results are then transferred to the correct locations on the numbered wheels ('addresses', in modern computer jargon). These operations are all under the control of the program.

If we could follow the operation of the machine on the number wheels, we would see the mill working (this is the origin of the expression 'number-crunching'). Then we would see the various wheels rotating, in turn, exposing new numbers, and returning to zero. The data values are first to move; then, in sequence, the wheels represent the products, then the subtractions, then the divisions.

Finally, the printer gives the answer. All wheels are now at rest, set at zero. The printer has given us the values of x and y. The machine is ready for the next set of equations, or for new cards if we wish to change the program. The machine rings a bell.

What we would see in the last problem would be the following succession of numbers:

Data read in:	1, 1, 60	2, −3, 0
Standing for:	$(m + w = 60)$	$(2m - 3w = 0)$
Products:	−180, 0, −3, 2	120, 0, 2, 0
Subtractions:	−5, −180, −120	
Divisions:	36, 24	
Printer:	$x = 36$	$y = 24$

(Plus and minus signs are shown, appearing and disappearing on special discs set above the number wheels.)

How the analytical machine was intended to work

The machine's secret is bound up with the fact that cogs and wheels engage together, one driving the other. The *direction* in which they rotate is an analogue for subtraction and addition. One direction of rotation adds, the other subtracts. Remembering that Babbage invented the odometer to measure distances travelled by railway trains (on a principle later adopted to measure the distance and speed of cars and other wheeled vehicles), we can see how the gears and cogs worked. We can imagine the train moving along, turning a cog wheel which engages with another cog wheel. So long as the wheels turn in the same direction, the operation is addition. The sum of the numbers is shown on the milometer. This is simply a set of wheels with numbers which are exposed in sequence until the motion stops. If the train is now made to go in reverse, the cog wheels will turn in the opposite direction, and the record will run in reverse. In other words, to add, the wheels turn in one direction; to subtract, the wheels turn in the opposite direction.

The principles so far described are basic only. To design a machine that could carry out any mathematical calculation whatsoever, without error, and print out the answer, took Babbage profound analysis over 40 years. One of his primary objectives was to improve the speed of calculation. Like the Egyptians, he used the fact that the operations of subtraction, multiplication and division can all be made special cases of the process of addition: to multiply is to add the same number repeatedly; to subtract is the same as adding the complement of the number being subtracted; to divide is the same as subtracting the divisor repeatedly.

The basic problem in all this, for a machine just as for a human new to calculation, is with 'carries'. When we add, sometimes the number increases to the point where it jumps from one level to the next: 9 units become one ten, or 9 tens plus 9 units (99) change the sum to one hundred. The other side of the coin is 'paybacks'. These arise when we have to borrow from a higher level in order to subtract a large number from a small one. When we subtract 8 from 3, for example, we need to borrow 10 and add it to 3; then subtract 8 from 13. Then we must pay back the 10 before we can proceed.

Babbage, adapting the ratchet idea from Leibniz's calculator, improved on it, devising the method of simultaneous 'carries' and 'paybacks'. To allow for a 'carry', the ratchet device moves the next cog in line forward one space. Babbage's method allowed for all numbers in a sum to be added or subtracted at once without doing the individual 'carries' and 'paybacks'. The correction for all 'borrows' and 'paybacks' is made in a single step. This brilliant idea speeded up the work-rate of his machine enormously: the difference in speed of the two procedures is of the order of 20 times.

When a human being adds or subtracts, it is a question of operating with two numbers at a time. We remember to add an extra one for 'carries', or to subtract an extra one for 'paybacks'. Because we do 'carries' and 'paybacks'

'Carries' in addition	'Paybacks' in subtraction
2 378 561 923	2 378 561 923
+ 1 293 658 197	− 1 293 658 197
= 3 672 220 120	= 1 084 903 726
carries: ++ +++ +++	paybacks: - - --

in our heads, we seem to do each of them in a single step. Actually, we do eight 'carries' in the addition sum shown, and five 'paybacks' in the subtraction. So, in the addition, we go through 25 distinct operations. The subtraction takes 19 operations.

We could program the machine to add or subtract exactly as we do. But this would entail many superfluous operations, of which we are normally unaware, with a corresponding waste of time. The machine can be programmed to do the same work in an entirely different way. It 'decides' first, comparing each number with the one beneath it, whether there is a 'carry' in the addition sum, or a 'payback' in the subtraction sum. This is step one.

Next, the machine takes each number and the one beneath it, and adds or subtracts them as the sum indicates. At this point, no 'carries' or 'paybacks' are made. All debts will be paid, and all credits collected, in one single operation. At this point, before this adjustment, it is simply a matter of recording the answer to each pair-by-pair add or subtract, forgetting about the next place in line. In subtraction, we borrow when necessary, but do not pay back. In adding, we do not add the carries. This is step two.

In the final step, the machine adds all the 'carries' to the individual total sums, or subtracts all the 'paybacks' from the individual differences. It is like the Jewish Festival Day of Jubilee when all debts had to be paid. It works as follows:

Addition		Subtraction	
2378561923		2378561923	
1293658197		1293658197	
**cccccccc		**p*p*p*pp	
+ 11111111	Step 1	− 010101011	Step 1
3561119010	Step 2	1185913836	Step 2
3672220120		1084903726	

*means do nothing; p means payback; c means carry.
The line of carries and paybacks is moved a space left (which is how the machine would 'see' it).

Although this takes a long time to explain, it is easy to see how much machine time is saved by using this algorithm. Ten numbers are added or subtracted in one operation, instead of one at a time. Almost 90 per cent of multiplication or division time is saved.

It was to do work like this that Babbage designed his analytical machine in place of the difference engine. The difference engine was too slow. It could only do straight runs, performing only one kind of operation. The analytical engine, by contrast, could be programmed to do everything the difference machine could do, and much more besides. Ada Augusta suggested that it might be taught to write music. Babbage worked out how to teach it to play noughts and crosses (tic-tac-toe), and hoped to teach it chess.

Babbage's achievement

Babbage worked as a consulting engineer and was highly successful as an inventor. His work was done mainly on problems connected with rail transportation and signals. But he was actually a trained mathematician, with a gift for applying his knowledge to real-life problems. He himself said that the thing he regretted most was that he was drawn away from the mathematical theory of functions into a life-long obsession with the building of the analytical engine. He solved all the design problems for the machine, so much so that it was built by his son, and ran successfully – but sadly not until after his death.

Whatever Babbage's own opinion of the direction of his work, later generations are greatly in his debt. Probably no contribution he might have made to the theory of functions could match his trail-blazing work in machining parts, preparing plans which set new standards in draftsmanship, and in his designs for the computer.

Babbage's efforts to design an all-purpose calculating machine led him, step by step, to the conception of the modern computer. In 1950, John von Neumann published the results of his discussions with Eckert and Mauchly, who had just invented the first American electronic computer. His report, however notable, in fact contains few advances, in principle at least, on Babbage's ideas of a century before.

Babbage made two noteworthy advances in machine language. First, he made it possible to give commands to the machine which enable it to process numbers. Provided that a mathematical or logical problem can be stated clearly and broken down into small, defined stages, a machine can solve it. It can perform operations recursively (that is, use numbers which it has itself generated in an earlier stage, then go back to the beginning and do

Babbage's analytical engine	*von Neumann's design*

The following are different names for the same things:

1. 'Store'	'Memory'
2. 'Mill'	'Arithmetic unit'
3. Transfer mechanisms	'Control unit'

4. Input and output devices are a feature of both.

The chief differences between the two machines are:

5. Uses decimal numbers	Works in binary system
6. Mechanical parts move	Electrons flow

the same calculation over again with the new numbers). It is also able to compare two numbers, then make decisions on the basis that the numbers are equal, or that one is bigger than the other. (The decision is usually about whether to continue, or to jump to some other part of the program, or to return to a previous jump-off point, or to give up.) It is this decision-making ability which gives the computer its power: the machine is behaving in a way once thought to be exclusively human and thus species-bound.

Second, Babbage invented a symbolic language by which to describe the structure and behaviour of the machine. This laid the basis for, and gave a clear direction to, the invention of symbolic logic. This new branch of knowledge was opened up by Augustus de Morgan, George Boole and several other 19th-century mathematicians. Babbage, by his early rocking of the boat and symbolic killing of the sacred cow of English mathematics, laid the foundations for a new beginning.

But none of Babbage's other achievements eclipse his position as founding father of the computer age. There is an historic link between his work and American innovations in this field. Just after the First World War, a New Zealand mathematician, Leslie J. Comrie, a 20th-century disciple of Babbage, became head of the Nautical Almanac Office at Greenwich, in London. He used a Babbage-type difference engine to compute the motions of the moon from 1919 to 2000. The data were 20,000,000 numbers, which he had had punched on half a million cards. In 1928, in the middle of his work, an American astronomer, Dr Brown, who had made the original astronomical observations, paid him a courtesy visit. Brown had worked on the calculations himself, by hand. As he admitted, he was in 'ecstasies of rapture' when he saw this machine churning out calculations with his numbers at the rate of 20 or 30 cards per minute.

The enthusiasm with which Brown described this experience on his return to the USA stimulated W.J. Eckert, a leading pioneer of electronic computers in North America, to continue work on his machine, assisted by J.W. Mauchly. This machine was eventually completed at the Moore School of Engineering in Pennsylvania. Eckert and Mauchly, in turn, passed on the torch of inspiration to John von Neumann, who built the computer at Princeton, and to Turing and Wilkes at Cambridge, the other main creators of the modern electronic computer. The chain of inspiration is clear, and it begins with Babbage.

1 2 3 4 5 6 7 8 9 10 11 12 13 14 15 16 17 18

BOOLE AND BOOLEAN LOGIC

'Who shaved the barber, the barber, the barber?
Who shaved the barber? The barber shaved himsel'.'
*Glasgow children's street song, 1920s**

SOCRATES: What Plato is about to say is false.
PLATO: What Socrates has just said is true.

George Boole (1815–64)

Boole was born in the cathedral city of Lincoln. His father was a self-employed carpenter. Some years before Booles birth (1815 was the year of Waterloo), Napoleon had contemptuously described the English as 'a nation of shopkeeper's. This was nothing to the contempt of the English upper classes for the group to which Boole belonged. To be a working man or woman then, even a self-employed tradesperson, was as bad as being a millhand or an Indian 'untouchable'.

This snobbery had a crucial influence on Boole; it was a key motive for his drive for self-improvement. His birth was 'unfortunate' only because of the obstacles it placed in the way of his undoubted genius. Little is known about his father except that he was interested in science and in his son's education. He shared his hobby with the boy; the making of scientific instruments, such as telescopes and magnifying glasses. He gave his son lessons in elementary mathematics, and arranged for him to learn Latin from a friend.

The condition of English education at this time was dire indeed,

---

*The words of this song parallel Russell's paradox that in a certain village the barber shaved everybody who did not shave himself. Logically, he could not both shave himself and also be one of the people who 'did not shave himself', i.e. those entitled to be shaved by him. In Glasgow, children such as myself cut this silly Gordian knot.

especially for women and the children of the poor. Reading was taught in charity schools to enable students to read the Bible. Writing and mathematics were absolutely forbidden in schools supported by Anglicans. They were considered 'gentlemanly pursuits' which would raise the aspirations of the poor 'above their station'. England had no state-supported schools until Parliament in 1832 provided the sum of £30,000 from the Exchequer (less than was spent on the King's stable in that year).

University education was little better. The English universities were monopolised by Church of England Tories, all other would-be entrants being excluded by religious tests. The education provided was tailored to the needs of huntin'-shootin'-and-fishin' country parsons, usually younger sons who had no money, no brains and no inclination for the army.

Boole broke through this élitist and inadequate system, first by educating himself, then by being selected as a pupil at a local 'normal school', a charity institution to select and train children of the poor as teachers for the other charity schools. He stayed there until he was 16 – itself an achievement in days when poor children often worked from the age of 8 or 9, and few stayed at school into their teens. He then took a position as 'usher' in a private school. While at the charity school, he had taught himself Latin and Greek; now he studied French, German and Italian so as to be able to read the continental writers on mathematics. These were intellectual achievements of the first order. In adult life, he continued to work as a teacher. He also served as a volunteer librarian at the Lincoln Mechanics' Institute. Such institutes were voluntary organisations, usually sponsored by local manu-facturers. Their purpose was to educate adult manual workers in the arts and sciences, especially as applied in industry and commerce.

As librarian, Boole had access to books and journals donated by the local businessmen who sponsored the Institute. At 25 he was the main support of his parents. He decided to open a day school, and to teach (among other things) the rudiments of mathematics. Consequently, he began to study the subject in greater depth. Like Babbage, Peacock, Herschel and de Morgan (founders of the new wave in mathematics which was centred in Cambridge), he was unimpressed by the standard of the textbooks available in English. He decided to use the language-skills he had so laboriously acquired, and to go to the fountainhead in his subject, namely the works of Gauss, Laplace, Lagrange, Leibniz, Euler and other continental writers. He carried out some original work in albegra, discovering invariants.

In those days, it was difficult to publish original mathematical work in Britain. However, a new journal had been started by the Cambridge mathematical faculty. Boole sent in a paper on the calculus; it was accepted by the editor, D.F. Gregory, a young Scot of Boole's own age who became a close friend. Altogether, while still teaching in his own elementary school, Boole contributed 24 major articles on mathematics to Gregory's journal, mostly on the calculus. As a result, he became well known in the tiny circle of English scholars doing innovative work in algebra and mathematical logic.

Gregory, who was a Fellow of Trinity College, Cambridge, devised a plan for Boole to enter the College as an undergraduate to study mathematics. This would have opened the way for him to take up a career more suited to his abilities. But it would also have involved him in giving up research for three or four years to work for examinations, not to mention abandoning responsibility for his parents' support. He refused the offer, stayed in Lincoln, and continued to labour in his own vineyard for almost ten years. He was a strong supporter of the Mechanics' Institute movement, but also involved himself in social problems affecting the working class. He was vice-president of the local Early Closing Association (to shorten the working day for shop assistants), and was a trustee of the local Female Penitents' Home (a voluntary society for the support of 'fallen women').

All this time, Boole's main research interest was higher mathematics. In 1844 the Royal Society awarded him a gold medal for his articles on the calculus; soon afterwards he began what was to be his major contribution to the subject, his work in the field of mathematical logic.

The basic idea that Boole worked on was one which the Cambridge school had earlier adopted. First declared by Charles Babbage in an unpublished work, and elaborated in print by George Peacock, it was the notion that mathematics was merely a particular case of a more general system of logical thought. This logic concerned itself with the relations between things and, especially, between classes of things. The relations could be numerical, or of other kinds. Boole's contribution was to devise a calculus (a method of representation by symbols) by which statements defining such concepts as 'wealth' or 'unclean beasts', as well as factual statements such as 'All human beings are mortal', could be translated into mathematical equations and manipulated using ordinary algebra.

As an illustration of his method we can take the proposition stated by the economist, Senior, as the definition of wealth: 'Wealth consists of things transferable, limited in supply, and either productive of pleasure or preventive of pain'. Boole deals with this proposition by first getting rid of the conjunction 'and', and then by introducing symbols: $w = $ wealth, $t = $ things transferable, $s = $ limited in supply, $p = $ productive of pleasure, and $r = $ preventive of pain. Then he is able to substitute an equation for the original statement:

$$w = st[p(1-r) + (1-p)]$$

All the operations of language, as an instrument of reasoning, can be represented by Boole's system of signs. It closely parallels the system of algebra. Boolean logic can be applied to any statement at all, provided it can be dealt with in terms of a binary classification: that is, put into binary form (true/false, belonging/not belonging, male/female, and so on).

In 1849, in spite of Boole's lack of a degree or special training, he was appointed to the Chair of Mathematics at the recently-established Queen's College, Cork, in Ireland. In 1854 he published his masterwork, *The Laws of Thought*, setting out rules for the system of manipulating symbols which in mathematics, is known as set theory, or Boolean algebra.

Boole died in 1864, aged 49. He was one of those gifted amateurs, outside the mainstream, who so often take tremendous strides forward in their field. He owed little, if anything, to traditional scientific wisdom – and indeed his work was neglected for some 50 years by leading mathematicians as having no significance. It was not until the 20th century that it came into its own. In 1937 Claude Shannon, an American research student at the Massachusetts Institute of Technology, first saw the connection between electronic circuits and Boolean algebra. Thus, Boole's work, interpreted by Shannon, provided one of the most basic half-dozen ideas in the computer revolution.

Boolean logic and the computer

In fact, the binary number system used by computers clearly demonstrates how Boole's binary logic works. Computers use the binary system because they operate by means of 'switches' which can be either 'on' or 'off'. They use two numbers only; 1 ('switch ON') and 0 ('switch OFF'). (The human operator works in decimal numbers, but the computer converts these into binary before using them.)

The table on page 213 shows the decimal numbers from 0 to 9 with their binary equivalents. It also shows the complement, or NOT, for each number. To create a NOT in binary, each 'bit' (individual numeral, 1 or 0) in the original number is changed to its opposite: each 1 becomes 0 and each 0 becomes 1.

The binary system reads from right to left, place values and numbers getting larger as we move to the left. This is the same principle as in the decimal system, where we start on the extreme right with units, then tens, then hundreds, and so on. In the binary system, instead of going up in powers of 10 as we move left, the place value goes up in powers of 2: in other words, each numeral (that is, each 1 or 0) is multiplied by the power of 2 that corresponds to its place in the number. For example, in decimal numbers, reading from right to left, 123 means $(1 \times 3) + (1 \times 20) + (1 \times 100) = 123$ (one

Decimal X	Binary X	Binary NOT X	In decimal
0	0000 0000	1111 1111	−1
1	0000 0001	1111 1110	−2
2	0000 0010	1111 1101	−3
3	0000 0011	1111 1100	−4
4	0000 0100	1111 1011	−5
5	0000 0101	1111 1010	−6
6	0000 0110	1111 1001	−7
7	0000 0111	1111 1000	−8
8	0000 1000	1111 0111	−9
9	0000 1001	1111 0110	−10

Note: For easy reading, throughout, each 'byte' (set of eight bits) is broken into two 'nibbles', each four bits long. This luxury is there only to help the human eye; the computer does not need it.

hundred and twenty-three). In binary, also reading from right to left, 111 means $(1 \times 1) + (1 \times 2) + (1 \times 4) = 7$ (decimal).

In the binary system, as in decimal, we multiply each place value by whatever numeral is in the place. In decimal numbers, this may be any number from 0 to 9; in binary it is either 0 or 1. We then add all the partial products. It looks like this:

Place number	8	7	6	5	4	3	2	1
Binary number	1	1	1	1	1	1	1	1
Value of numeral =	2^7	2^6	2^5	2^4	2^3	2^2	2^1	2^0
	= 128 +	64 +	32 +	16 +	8 +	4 +	2 +	1
	= 255							

NOT-ing, AND-ing and OR-ing

In our normal way of talking, the word 'AND' is taken to mean 'together with'; 'OR' means 'one but not both'; 'NOT' means 'the opposite of'. In Boolean, binary logic, the words are 'logical operators', and mean something radically different.

A good way to understand how these logical operators work is to compare them with electric or electronic circuits. Each logical operator (NOT, AND, OR) can be identified with a particular circuit.

Suppose we have a current that flows along circuits through a set of switches. The switches act independently, so that any or all of them can be either ON or OFF. (These states are represented as 1 or 0 respectively. If two or more switches are connected in series (forming a single path), then

Logical AND	Inclusive OR
Series Circuit, all	Parallel Circuit,
switches ON	one switch ON, one OFF
	SW(A)
—SW—SW→	→!off!
(A) (B)	→SW(B)→

all must be ON to permit the current to flow. One switch AND all the others must be ON. If two switches are in parallel, they can provide two paths for the current. The one switch, OR the other, OR both can be closed to make a path. The current flows in a parallel circuit if one OR other, OR both switches are ON. This can be shown in a table, as above.

When more than one switch in a parallel circuit is ON, the current divides, each fraction flowing along a different path to re-combine before leaving the circuit. In binary logic we are only interested in whether the current flows (a logical 1) or does not (a logical 0). The size of the current is of no interest. There are three essential principles: first, the logical operators AND, OR, NOT and so on mean something different from their normal usage; second, the operators act on binary numbers only; third, the operations of NOT-ing, AND-ing and OR-ing are carried out *on each bit* of the binary number separately.

Another way to understand the logical operators is by means of truth tables. These show inputs and outputs for the circuits corresponding to each operator. For example, suppose we have two switches A and B. We will again use 1 for ON and 0 for OFF. The tables are called 'truth tables' because 1 represents 'true' and 0 represents 'false'.

An AND situation is a series circuit; an OR situation is a parallel circuit. In a NOT circuit we have an inverter which reverses the state of the current – it flows only when the switch is OFF (0). The truth tables for these three cases are given below:

AND (series circuit)		OR (parallel circuit)		NOR (NOT OR) (inverter)		
A B	A AND B	A B	A OR B	A B	NOT (A OR B)	
$1 \times 1 = 1$	T	$1 + 1 = 1$	T	1 1	0	F
$1 \times 0 = 0$	F	$1 + 0 = 1$	T	1 0	0	F
$0 \times 1 = 0$	F	$0 + 1 = 1$	T	0 1	0	F
$0 \times 0 = 0$	F	$0 + 0 = 0$	F	0 0	1	T

For switches and output, 1 is ON, 0 is OFF.
The NOT OR circuit is called a NOR; a NOT AND is a NAND.

On large computers, equipped to support Boolean operators (as some home computers are not), other operations are also possible. In fact these can all be mimicked, by using combinations of AND and OR circuits, with or without extra inverters.

Exclusive OR-ing, IMP-ing and EQV-ing

In large computers, three other logical operators are often built in. Instead of 'either/or, or both', as in the inclusive OR (described above), the exclusive OR, or XOR, means 'either/or, but not both'. In this, with both switches ON, no current flows. The EQV operator is the logical 'equivalent' statement. It corresponds to the statement: 'p implies q and q implies p'. Here, if A and B are the switches, current flows only when A and B are the same, that is both ON or both OFF. The IMP operator is the implication operator. Here the current flows except when A is ON and B is OFF. This operator corresponds to the case stated in logic as 'if p, then q'. The electronic model is an OR circuit with the switch A acting in reverse.

The importance of Boolean logic

Computers will come more and more to dominate our culture. Boole's logic is a purely abstract, intellectual system, designed before Babbage completed his computers. It was conceived as a way to represent the language of logic in symbols; Boole had no inkling of any practical application for it. But in 1937, when Claude Shannon showed that it was a tool that could be used to reduce the number of components needed to make a computer, he cut down at a stroke the number of circuits, the cost of the whole enterprise, and, most importantly, the amount of heat generated inside the computer and causing breakdowns.

Computers depend on Boolean algebra to do their work. Boolean operators are 'burned into' the hardware so that the computer can solve logic problems. Whether or not Boole's work is comprehensible outside the expert fields of mathematical logic and computer mathematics, his thinking, sidelined for almost a century, is essential to them both.

1 2 3 4 5 6 7 8 9 10 11 12 13 14 15 16 17 18

MACHINES WHICH (OR WHO?) THINK

> Professor von Neumann has proved that in theory machines can reproduce themselves, but no one has yet explained this to a machine, and so far, they have never shown any signs of taking the initiative.
>
> *B.T. Bowden, 1953*

Alan Mathison Turing (1912–54)

Through his mother, a major influence in his life, Turing was descended from an old Irish family, the Stoneys, which included many distinguished inventors, engineers and scientists. His grandfather on his father's side had been a student of mathematics at Newton's and Babbage's old college, Trinity College, Cambridge. As an undergraduate, he had the notable habit of sleepwalking on the roof, but survived to become chaplain of the College. When Turing himself graduated in mathematics, he became a Fellow of the college.

Turing showed early signs of being out of the ordinary. Even as an infant, he showed great interest in numbers. At the age of three, he learned to read in three weeks. Later, one of his teachers recognised that, although he manifested what was called an 'unclubbable personality', he also had a 'periscopic mind': that is, able to see an answer long before others had even grasped that there was a question. In his lifetime, Turing constantly showed the two abilities he himself identified as necessary for a good mathematician – ingenuity and intuition.

Turing's most important contribution was a paper he wrote in 1936 on 'computable numbers'. Here he showed that some numbers cannot be worked out by any definite, fixed process. (This was a kind of corollary or restatement of Kurt Gödel's (1906–78) theorem, which demonstrated, similarly, that mathematics as a science can never be totally consistent and totally complete.) It was the solution to a well-worn problem and was not terribly interesting, except perhaps to mathematicians. What was new or important was the test that he invented for a 'fixed and definite process'. He defined this as a routine which could be carried out efficiently by a machine. In other words, he was saying that if a number cannot be calculated by a computer such as Babbage's machine, then it just cannot be calculated at all.

This may sound like a cop-out, or even a play on words, but Turing took pains to describe what he meant by a machine, and his description became part of the language: we now talk about a 'Turing machine'. This was not just a talking point (though his words have generated hundreds of thousands more); he went on to design and make one of the first electronic computers, while working in Manchester. It was called the Ace.

We have to remember that, in the late 1930s, the only 'thinking' machines in existence were calculators that worked on purely mechanical principles. They were a matter of wheels and cogs; they depended on gravity, muscle power, springs or electricity to drive the mechanism. The analytical engine projected by Babbage was the highest peak of achievement in this mode. The Turing machine was a completely new idea, a breakthrough in predicting the next level of 'thinking' machinery. It reverted to first principles. Euclid had defined a number as an aggregate made by adding one unit to another as often as was necessary. Turing put this idea to work in a totally new context.

The most important feature of the Turing machine is that it not only handles numbers but can be extended to anything human beings can do. So long as we can break the operation down into stages, and state what these are without any doubt or ambiguity ('a fixed and definite process'), then a machine can be devised to replace the human operator. This covers a lot of ground. It starts from a totally new idea of what machines are capable of doing. We would now recognise as Turing machines such items as a washing machine, a word processor, a general or all-purpose computer, or a robot in a car factory. These can all be programmed to carry out 'fixed and definite' processes, but, of course don't answer back.

Turing was one of the world's eccentrics. He liked to do everything for himself. Day-to-day problems were taken as a challenge to his ingenuity. Sometimes his solutions were bizarre in the extreme. For example, he had to cycle daily to Bletchley, where he had about 100 female civil servants as clerical assistants. He wore his civilian respirator (gas-mask) *en route*, to overcome attacks of hay-fever to which he was prone. In the winter, he wore woollen gloves that he had knitted himself without the aid of a pattern. He had taught himself to knit, but never found out how to finish off the fingers, so the gloves had long pieces of wool hanging off them. As he had a problem with his bicycle chain (which came off after a certain number of revolutions of the wheels), he would be counting the number of pedal strokes he made. According to a formula he had worked out, the chain had to be adjusted after so many hundreds of revolutions. So he would stop just before then, take off his gloves, adjust the chain and proceed on his journey to work. His mother commented privately that five minutes and a good mechanic would have permanently solved the chain problem.

One of Turing's major interests was exercise. He liked to go on country marathons, dressed in garden trousers tied up with a piece of rope. Since he needed to plan these runs and his watch was faulty, he used to tie an alarm clock on the end of the rope. When visiting friends in Cambridge or London, from Bletchley, he would often send his 'decent clothes' in advance and run 25 or so miles to their home. Or he would run similar distances home after a party. He was in the Olympic class as a marathon runner, and a member of the same club as Roger Bannister, who later ran the first four-minute mile.

There was method, as well as independent thought, in everything Turing did. He had very little regard for convention, either for its solutions or for its appraisal of his behaviour. In the end, this habit of doing everything for himself was literally the death of him. For years, he had been running chemical and electrical experiments, using home-made apparatus and materials. His plan was to discover all the chemicals that could be made from ordinary kitchen or household materials. Some of these experiments were quite exotic. For example, he went in for electroplating, using home-made batteries and his grandfather's gold watch, to plate some kitchen spoons. (This could account for the alarm clock as an adjunct to long-distance running.) One of the salts he used in these electroplating experiments was potassium cyanide. He also had the rather childish habit of touching things as his experiments proceeded. On this occasion, he must have handled one of the spoons, and rubbed some cyanide off on his fingers without noticing. He was discovered dead in bed one morning in June 1954.

The Enigma project

When war broke out in 1939, Turing was recruited to work on the Enigma project. This was highly secret work on a coding machine to be used to decipher enemy signals. At this time, the German High Command was using a machine which, starting from a normally-typed message, produced a coded translation automatically. Since the process was completely automatic, the code could be changed very easily. All that was needed was to set up the code and type in the message, which was then sent in code by teleprinter and decoded automatically by the machine at the other end.

Using such a machine, it was possible to produce messages that could not be readily deciphered by traditional methods.

With the help of a Polish engineer, Richard Lewinski (who had worked on the prototype for the Nazis until they discovered that he was Jewish, and who was smuggled out of Warsaw in 1938), the British made a copy of this machine. Turing's job was to make a mock-up, and try to decode the signals from the German High Command sent daily on known wavelengths. Thus he would discover by trial and error, from an infinite number of possibilities, the code for the day. The German machine was called Enigma by the British. (This was an in-joke, referring to the number of 'variations' produced by the machine, an allusion to the *Enigma Variations* by the English composer Elgar.) The German machine was, in fact, a Turing machine; Turing was clearly the right person to work on it.

It consisted of two typewriters, one located at headquarters, the other in the field. Each had an extra part: what was then called a black box. (This expression was used by wartime scientists to describe an appliance into which you fed an 'input' and from which you received an 'output'. The layperson – a corporal in the field, say – could use it without knowing, or needing to understand, anything of what went on inside.) The black box here was the Enigma machine, which first scrambled and then sorted out given messages.

Turing's brief was to break into the German system: to devise a machine which, without 'knowing' the codes in advance, would take the intercepted message and automatically work out the code. He probably used a recursive algorithm like Newton's trick for working out the square root of a number (see page 183). Here, you guess the answer to begin with. The machine then calculates the square of the guessed answer and compares this with the original number. You can then estimate your error and adjust your first guess upwards or downwards. By this procedure, repeating your guesses and correcting them each time, calculation gives you an improved result. You correct your answer at each step so that it gets closer and closer to the real value as you go on. The virtue of this procedure is that you can start with any guess whatsoever, since it is a self-correcting process. It very quickly converges to the correct answer. It is relatively easy to write a program to do this on a modern table-top computer.

The same thing happens with the code. In the interest of economy of time, you guess the most likely solution. The machine is not interested in how wrong you may be. It zooms in on the translation in a progressive way, using the code. This is modified by the machine, which very quickly makes more and more sense of the message. In the exact sense, Turing's task was to invent the first 'word cruncher'.

As the Nazis made their machine more and more massive to accommodate more and more abstruse codes, so the British were driven to invent bigger and bigger electronic decoders. By December 1943, Turing had invented the world's first electronic special-purpose computer. It was known as Colossus. It had 1800 thermionic valves (in the form of small vacuum tubes) and was fed with paper tape punched with patterns of holes, the different patterns standing for letters of the alphabet. The coded message was fed into the machine on one tape, and the program with the decoding instructions on another. The tapes were punched using the teletype code, each letter consisting of five holes punched in different positions. The message and the program were brought together inside the machine by a third piece of tape known as the 'job description'. This stated the date and time of the message, the nature of the operation to be carried out (instructions for decoding), and any other relevant details.

Once the machine was set up, it would then do its work at an unimaginable, 'colossal' speed. For example, it could read characters from tape at the rate of 5000 per second by means of photoelectric cells. It could do all the tests (that is, substitutions) at about a million per second.

The Turing machine

In 1936, when Turing wrote his account of computable numbers, the Turing machine was merely a hypothetical, abstract model – a kind of fantasy, or piece of science fiction. Turing was actually a mathematical logician, like Bertrand Russell, whom he very much admired. In describing a 'fixed and definite process', such as solving an equation in algebra or calculating a square root, he captured the essence of an algorithm. Basically, what Turing was saying was that, in these simple cases – in fact, in any problem that we can solve – we can spell out a set of routine procedures which, as required, can, in principle, be carried out by a machine. All we need to do is to break the problem down into simple steps, work out for each step clear instructions for the machine to follow, and then stand back. We can be sure that if the machine cannot solve the problem, then it cannot be solved at all. (For example, there are some numbers whose nature is such that they cannot be calculated, for example, computers can't divide by zero.)

Reduced to its simplest elements, a Turing machine is one that can 'read' (take in) simple instructions, written in some suitable code. It is also able to follow and carry out the instructions, which are broken down into very simple steps and presented to it one at a time. It must be able to go back and change its earlier 'read' if this becomes necessary and to take up where it left off. It must also be able to stop after completing its task and to do nothing while waiting for the next task. Washing machines don't really qualify.

In its application to computing, a Turing machine is an abstract device, consisting of three parts. There is, first of all, a control unit, which can accept and pass on a number of instructions or commands. There is, next, an endless tape (one that can be added to repeatedly, before it runs out), marked out in unit squares. Each square can be written on by the machine; the machine can also read what is written on each square. Only one piece of information, a number or a letter, can be written on each square. The square may be blank in some cases, and if necessary the data can be changed by moving the tape back and writing over what is on the square. The tape can move, right or left, backwards or forwards, one square at a time. The third element is the read–write head, under which the tape passes, square by square. The single symbol, written on each square, can be read and passed along to the control unit. It can be either a single piece of data, or a code letter or number standing for a command. The behaviour of the machine is controlled at all times, first, by its internal state, which is a product of its earlier activities; and second, by the new data or commands passed to it from the read–write head.

The computer as counsellor

Another question posed by Turing was, can we make a machine that can respond to questions in such a way that it is impossible to tell that it is a computer and not a human being responding? We are now in a position to answer this. We can easily construct a dialogue, for example, between a simulated Rogerian counsellor (one who gives psychiatric counselling on the principles of Carl Rogers) and an imaginary client. For the purposes of comparison, the first dialogue below is an excerpt from a real interview.

The second dialogue is produced by a program where the 'client' types a question for the computer 'counsellor', or answers a question posed by the same, or makes a statement. Key words are singled out from this input as signals for the possible replies. A prepared reply is chosen from the program, by a random process, from a number of options which include relevant responses. The computer options are 'set pieces', whereas the 'client' is free to say anything (that is, to type it in as his or her part of the interview). As in Rogerian counselling with a live counsellor, the main function of the computer-counsellor is to keep the client talking in an emotionally supportive atmosphere. It is described as non-directive therapy, because there is no diagnosis or treatment plan. Quite simply, the patient decides what to say and this is accepted unconditionally by the counsellor in a positive, supportive way.

Counselling session A: Rogers with Gloria

Therapist: Good morning. I'm Dr Rogers. You must be Gloria.

Client: Yes, that's me.

Therapist: Good. Now tell me about your problem.

Client: Well, right now I feel very nervous, but I don't think you'll be too hard on me.

Therapist: Um-hum, um-hum.

Client: The problem is that I'm recently divorced and I have an eleven-year-old daughter. What bothers me is that I sometimes have men to the house and I wonder how this affects the children, especially my daughter. She's old enough to question whether it's right for me to do these things.

Therapist: I wish I could give you the answer about what you should tell her.

Client (laughing): I was afraid you'd say that. What I really want to know is how it would affect her if she found out that I was lying to her.

Therapist: Yes. You mean she'll suspect something's not quite right?

Client: Then there's the fact that I feel guilty when I talk to her about this. I want her to accept me.

Therapist: Now I can hear what you are saying. What you'd like is to feel more accepting of yourself. This is a pretty tough assignment.

Counselling Session B: (a fictitious client)

Therapist: Good morning. I'm Dr Rogers. It is Gloria, isn't it? I believe you have a problem.

Client: Yes, doctor. It's my live-in companion. He's taken to beating me for no reason at all. Can you help me?

Therapist: Can you tell me more? What's the nature of the psychological problem?

Client: Well, I'm recently divorced, and I have a live-in boy friend. He's given up his job and depends on me for food and everything. He drinks and sponges on me for money. I put up with this because I love him, but recently he's taken to beating me for no cause at all. I feel guilty about this because I didn't make a stand at the beginning and now things are going from bad to worse.

Therapist: What does this suggest to you?

Client: Well, my main worry is that I find it impossible to break out of this vicious circle. I don't seem to have any will-power.

Therapist: Tell me more about your feelings.

. . . *and so on.*

The non-directive, non-judgemental nature of the therapy in Session B is brought out by extracting the therapist's contributions. These can be seen as having no specific reference to Gloria's problem (the therapist avoids all specifics) but are of such a character that they could be said in any therapy situation. All this testifies to the validity of Turing's view that it is quite possible for a machine to duplicate human (verbal) responses to complex situations in such a way that the responses cannot be distinguished from those of a human being.

Machines with feelings?

Another question posed by Turing, most difficult and most significant, remains. Could we make a machine that would have, or would manifest, the same feelings as human beings (or the feelings that human beings claim to have)? The short answer is, yes, provided that we abandon all our preconceived ideas about 'machines' and 'humans'. We need to be clear about the nature of computers and the nature of human feelings. It is relatively easy to understand the former; the difficulty is in obtaining consensus about the latter. Success in making a machine with feelings would, of course, drastically change the nature of the machine. In fact, some people would say that it had ceased to be a machine and had become a 'person'. As of now, the computer is an automaton that can simulate certain kinds of human mental skills, much in the way a washing-machine can simulate the physical behaviour of a human being. The washing-machine can go through all the motions of someone washing clothes – heating water, putting it in a tub, adding cleansing agent (liquid or powder), agitating the water, emptying it out, adding fresh water to rinse the clothes, and removing most of the water from the clothes by centrifuge and a current of warm air.

The only way this differs from a human being is that the washing machine, like all automata, absolutely depends on some human being to perform all the non-physical operations associated with the process. It is a human being who decides to initiate the process (by pushing a button, or by setting a time clock, etc.). It is a human being who monitors the process to make sure that the automaton does not get out of hand or break down. The human being is motivated to do all this to please someone (perhaps only himself or herself). When the operation is complete, it is the human being who has a sense of satisfaction or relief, or experiences some other emotion.

By its nature, the computer is simply a complex and elaborate automaton. It can solve a great number of complicated calculations that most people are incapable of doing, just as most people are incapable of washing clothes 24 hours a day for a week, say, without intermission. Nobody needs to be afraid of a (properly constructed and properly operated) washing machine. Similarly, there is no need to fear the computer. Like the washing machine, it is a mere extension of the human operator.

There is another similarity between the computer and the human being – one that really goes to the heart of the matter. The scientific (mathematical) model of the neuron or brain cell is the same as the model for an electrical relay. (A relay is a switch that reacts to changes in circuit conditions by carrying out one of a variety of operations with electric currents: see page 212.) The brain has a storage capacity thousands of times greater than the largest computer. On the other hand, the computer works several times faster than the brain. They both depend on particular kinds of electron flow as their source of energy. Both brain and computer are physical (electronic) systems, with a very similar overall organisation. The essential difference is that one is composed of living matter, the other of non-living matter. This means that there is an organic unity in the brain whereas in the computer we have simply an accumulation of parts. These are 'organised' in systems and subsystems, but there is no 'organic' unity.

In a living organism, there are processes of growth, healing and regeneration. Feelings are associated with living tissue even at the most primitive level. The simplest of organisms avoid danger and discomfort; they feel pain and respond to their own euphoric condition by approach behaviours. Even at the level of the cell, living matter is not passive. It is an active agent, food-seeking and philoprogenitive. All these processes are totally lacking within a physical system. The processes of growth, repair and regeneration are catered for in the physical system by support systems, such as trained technicians, depots for spare parts, a company organisation with records of purchases and repairs, tools, and so on. All these are inputs from outside the physical system.

The other essential feature of the organic system is that it can reproduce itself. Now, there is no doubt that we can simulate many aspects of living systems, including reproduction. (For example, we can set up automatic factories, controlled by robots, to manufacture other robots.) There are already laboratory 'factories' set up to make living cells and living tissue. To invent an automaton with feelings we have merely (merely?) to wed the potential of the miniaturised integrated circuit to the potential of the living cell. Given the means to structure such an automaton's environment, reinforcing the kinds of response we aim at, and extinguishing those responses we consider undesirable (we also need a large amount of time), there seems no reason why it would not be possible to produce a chimera made up of silicon chips as one major component and living, breathing, reproducing, passionate and self-active cells as the other. (One may well say, 'Then we would really be in trouble!')

No one knows exactly what Turing was aiming at in his uncompleted work on morphogenesis (the shape of living matter). On the assumption that this was not just a diversion from the main plan of his work, the most plausible explanation is that it was precisely this problem he was tackling. Just as his paper on 'computable numbers' revolutionised our perception of the machine, so he may have conceived the idea that his work on morphogenesis could provide the ultimate principles necessary for effecting a marriage between the computer and the living cell. Indeed, he made several radio talks on the themes mentioned above, with a similar emphasis to the one adopted here.

THE ELECTRONIC COMPUTER

'Explain my behaviour as I sit here saying: "No black scorpion is falling on this table".' *Alfred North Whitehead to B.F. Skinner. arguing against behaviourism at Harvard.*

OF the many themes running through the history of human thought, one of the most striking is the binary assumption (see page 1): that is, the notion that there are two opposite principles at work in the universe. These are variously described as the male and female principles, yang and yin, an excess and a deficit deviation from the 'golden mean', or a positive and a negative quantity. Some see these opposites as constantly seeking each other to reunite and reestablish the original unity. Plato, for example, thought of each human being's life as the record of his or her search for that other self (the alter ego) from which the individual was separated before or at birth – and said that when the two selves were reunited, the person would be free of all tensions and conflicts. Others, by contrast, see the two principles as being in eternal conflict. Hegel, for example, believed in a dialectic of conflict between them at the roots of reality. This conflict was creative and generative, the source of a new synthesis.

In the 17th century, when Gilbert and other scientists began to study magnetism seriously in connection with navigation by means of the compass, they accepted the Chinese concept of a twofold division into north-seeking and south-seeking poles. In the 19th century, a division into positive and negative electricity was soon made to account for what was being discovered.

Some 100 years ago there was a breakthrough in the study of the way matter is made up. The discovery of radioactivity by the Curies, of X-rays by Becquerel and of the electron by J.J. Thomson prepared the ground for new theories of matter and for Rutherford's model of the atom (1911). This is now the basis of our picture of the material universe. The experimental programme that lay behind this model was vast. But the most central experiment was carried out in J.J. Thomson's laboratory (the Cavendish) at Cambridge University, using a cathode-ray tube (invented by Braun in 1897). The earliest tube was a glass vessel from which the air had been expelled to form a vacuum. Inside the tube were two pieces of platinum connected to a source of electricity. These were the anode (positive pole) and cathode (negative pole).

When the current was switched on, the anode began to glow. It continued to do so when the current was off. If a solid object (a metal cross, for example) was placed between the anode and cathode, its silhouette became visible in the region of the anode. This suggested that something was passing from the cathode to the anode and was stopped by the metal object. Thomson found that these rays could be deflected by means of a magnet, and was able to show that they were made up of very small electrically charged particles. He identified them as 'electrons', which had already been proposed as one of the basic constituents of the atom. The 1890s thus marked the beginnings of the science of electronics.

Transistors

The transistor was invented by Brattain, Bardeen and Shockley in 1948. It inaugurated a new era in the manufacture of radio and television sets, and had an almost immediate effect in the computer field. The transistor is a device that replaces the cathode-ray tube and valves formerly used in radio and television. It produces electrons which are generated and flow through the silicon or germanium that the device is made of. Transistors could be made quite tiny and at practically zero cost. Their size was crucial for computer technology. Until their invention, electrons had been produced in thermionic tubes (valves) by wire filaments heated by electricity, or in cathode-ray tubes. With 10,000 or more such valves, large mainframe computers, because of the heat generated, were as often out of commission ('down'), for replacement of burned-out valves, as they were available for work ('up').

The modern digital computer is made up of a large number of transistors. The IBM microcomputer on which the original typescript for this book was produced, for example, has 655,360 bytes (eight binary digits: see page 213) in its working memory (RAM) alone, that is, 640×1024. (This is normally expressed as 640K; K stands for kilobyte, that is 1024 bytes.) This grand total of 5,242,880 ($8 \times 640 \times 1024$) transistor flip-flops (see page 231) and capacitors (which act as stores) is supplemented by a 'hard' disk and a 'floppy' disk, which make more than an additional 10 million bytes available for storing typescript and/or working with and storing numbers.

(The disks on which text copy can be saved, do not, of course operate as transistors.)

Nowadays the flip-flops are built in place on the 'motherboard', as an integrated circuit. The transistors, suitably adapted, function as electrical components such as coils, switches, resistors and capacitors. They serve the four computer functions: input–output; memory; arithmetic–logic unit; and the central control unit. (The discussion of structures and functions which follows, though general, is based on the IBM microcomputer, which established the 'industry standard' for these machines.)

Input–output

The input–output section of the computer is like an electric typewriter with some extra keys. Its purpose is to control and edit the flow of information between the keyboard operator and the computer. Information is displayed on a monitor (a cathode-ray tube, TV-like screen). This shows both the results of work done at the keyboard as it happens, and the results of earlier work that have been saved and stored in the memory (as an option).

The input–output sections act as a buffer between the high-speed computer and the much slower peripherals, such as the printer and (human) operator. The normal function of such sections has historically been to permit the operator to pose problems to and provide routines for the computer, to communicate with it and to receive results from it. This is done by 'loading' the computer with a program of commands, and then feeding it the data to be processed. The machine types the solution, following the sequence of instructions given in the program. Its report on the problem – its solution – appears as the 'output', whether on screen, printed on paper, or stored on disk or tape.

Memory

The computer memory corresponds to the 'store' in Babbage's design for the cogs-and-wheels prototype of the analytical machine (see page 197). In place of Babbage's discs (wheels engraved with the numbers 0 to 9 round the edges), transistors carry out the calculations. This work is done using 'bytes', the eight-digit numbers of the binary system. When words (as text, or instructions in programs) are used, they are converted to digits by the built-in ASCII conversion code (see page 163), before being stored in, or retrieved from, the memory in the usual way. (ASCII stands for American Standard Code for Information Interchange.) The fact that everything in the memory is numbers allows words and graphics to be manipulated step according to predetermined schemes or algorithms, in the same way as other numbers.

Arithmetic–logic unit

The arithmetic–logic central processing unit performs two kinds of operation, arithmetic and Boolean logical operations (see pages 213–15). The arithmetic unit carries out all calculations proposed by the program: not only addition, subtraction, division and multiplication, but also exponentiation (roots and powers). Tables of logarithms and trigonometric functions are also on call. Boolean logic functions assist this arithmetical work.

Central control unit

The control section oversees the automatic operation of the computer. For example, it 'reads' programs and sends signals to the various sections of the computer, organising the activities necessary to carry out the program. It separates instructions from data and sends them to different parts of the memory. Data words, converted to numbers, are sent to registers in the arithmetic unit; instruction words are sent to the instruction register in the control unit. The computer also has a precise timer, enabling it to synchronise the operations specified in the program, in hundredths of a second. This clock is vital to all control operations.

The register

About 90 per cent of the work of a computer consists of transferring data, in the form of number bytes, from one register to another. The register normally consists of eight locations (known as bits; the individual characters in groups of eight form bytes), but some computers operate with 4-bit registers, and others may have 16-bit registers or more. The registers store 'words' in the memory. Once again, the use of the expression 'word' can be confusing to non-specialists. It does not mean a word in the usual sense (such as the words on this page), but an ordered set of characters (numbers, not letters) that are stored, transported or otherwise treated as a unit. 'Words' can thus be sets of numerals coded to represent anything at all: figures, letters of the alphabet, continuous text, Japanese characters, printing commands in the form of graphics characters indicating printing formats, or characters such as space, back-space, new paragraph etc.

The A register, known as the accumulator, is the most important register in the computer. All calculations are done by moving the 'word' into this register, then transferring it to the central processing unit, performing the operation on it, and finally moving the changed quantity back to the accumulator and then to another register in the memory.

The method of storing, or anchoring, each bit (single character) in the memory is by means of a flip-flop. This is a device that remains either 'set' (at one) or 'reset' (to zero) until its state is again altered. It is like a switch (on or off) or gate (open or shut), whose state alters the electron flow and so affects the message travelling through the computer. To maintain its position (set or reset), each flip-flop is 'refreshed' from time to time, every one-thousandth of a second or so, by an electric charge.

Unlike flip-flops (which are 'bi-stable', able to be placed in one of two possible states), registers can be manipulated in a number of ways, altering the numbers originally placed in them. For example, a 'word' such as 0000 1111 (15) can be moved two places left, to become 0011 1100 (that is, 60). Or it can be moved right four places, converting it to 0000 0000 (zero). The various permutations of such movements enable the computer to process data (numbers).

The 'flag', or F register, is an 8-bit register whose eight individual flip-flops are set or reset at need by the arithmetic–logic unit as the various logical and arithmetic operations are carried out. The flags are really reports on the state of the accumulator or working area. As its name implies, each flag is a signal, that is, a one or a zero in the 'flag' register. This register, and other sensitive indicators, are systematically and regularly scanned (scrutinised) by the operating system. For example, if the accumulator is zero this is signalled to the control unit by 'hoisting' the zero flag: putting a 1 in the zero flag register.

The Z (or zero) flag is one of several. Others include the P flag (or parity flag, which signals whether the sum of bits in the accumulator is odd or even, confirming that no part of the number has been lost), the C and A/C flags (the carry and auxiliary carry flags, which signal whether addition and subtraction operations have been completed successfully) and the S flag (the sign flag, which signals that the accumulator is negative).

The control unit regularly reads and interprets flag signals as part of its 'rounds'. It monitors the changes signalled by the flags, overseeing and testing so that the computer can decide whether the order of processing is satisfactory, or needs to be altered to avoid a malfunction. All such checking is done automatically and systematically by the computer. The user can remain oblivious to the hoisting and striking of flags, without suffering any penalty.

The computer as number processor

Before describing how the computer works with numbers, we need to restate two facts. First, a computer does its work at almost unimaginable speed. A simple calculation, like adding two numbers, can be done in less than a millionth of a second. In timing or setting up computer operations, we are dealing in terms of the nanosecond (10^{-9} seconds; one thousand-millionth, that is, one US billionth, of a second). Second, it reduces all mathematical operations (including basic subtraction, multiplication and division) to a single process of addition. To find a parallel, we must travel back in time through the intellectual history of the human race. We must forget number skills mastered at school, and ignore most of the sophisticated counting methods devised over millennia to facilitate both abstract thinking about number and the exchange of goods and services. We must return, in essence, to the original operations with 'counters', performed by our remote ancestors in setting up hunting expeditions or sharing food. However arcane the workings of the computer may seem – and granted the difference in speed of operation – they are essentially no different from the simplest mathematical procedures devised by the human race.

Simple algorithms for the basic processes

How does the computer simplify the four arithmetical operations? Using the two numbers 10 and 5 as our examples, and expressing them first in decimal, the four processes are:

Addition	Subtraction	Division	Multiplication
10	10	$\frac{10}{5}$	10
+ 5	− 5	= 2	× 5
15	5		50

In binary, each of these numbers would appear as 8 bits, that is, one byte. We will show all binary numbers either as two sets of 4 bits (that is, two 'nibbles'), or as one nibble if this is sufficient. We now show the four procedures using binary representations of the numbers 10 and 5. (One nibble will do for each number, as the nibble can stand for numbers between 0000 and 1111, that is, in decimal, 0 and 15.)

(10) 1010	1010	1010	1010
(5) +0101	−0101	0101	×0101
1111	0101	=0010	110010
(15)	(5)	(2)	(50)

In the pre-literate state of mathematical knowledge, it is probable that sticks or stones were used as counters in the ways shown below (we use decimal numbers for convenience).

Addition	Subtraction	Division	Multiplication
10	10	10	10 one ten
+ 1 = 11	− 1 = 9	− 5 = one	+ 1,1,1, ... 10
+ 1 = 12	− 1 = 8	———five	
+ 1 = 13	− 1 = 7	5	= 20 or two tens
+ 1 = 14	− 1 = 6	− 1 = 4	+ 1,1,1, ... 10
+ 1 = 15	− 1 = 5	− 2 = 3	
		− 3 = 2	= 30 or three tens
10 + 5 = 15	10 − 5 = 5	− 4 = 1	+ 1,1,1, ... 10
		− 5 = none	
		———left	= 40 or four tens
		10 = two	+ 1,1,1, ... 10
		fives	
			= 50 or five tens

The same table in binary would show these number relations as follows:

Addition	Subtraction	Division	Multiplication
10 1010 = 10 ten	10 1010	1010 10	1010 1 ten
+ 1 = 1011 = 11	− 1 = 1001	− 0101 − 5	1010 2 tens
+ 1 = 1100 = 12	− 1 = 1000	———	1010 3 tens
+ 1 = 1101 = 13	− 1 = 0111	0101 − 1	1010 4 tens
+ 1 = 1110 = 14	− 1 = 0110	0100 − 2	1010 5 tens
+ 1 = 1111 = 15	− 1 = 0101	0011 − 3	———
		0010 − 4	11 0010 = 50
		0001 − 5	
10 + 5 = 15	10 − 5 = 5	0000	10 × 5 = 50
		10 = two fives	
		$\frac{10}{5} = 2$	

These binary routines are so tedious for the majority of people that they have been programmed and burned into the computer's 'read-only' memory (ROM). This means that they cannot be changed or wiped out. ROM is permanent. In the same way as with flags, the user can act as though totally ignorant of the binary system, merely telling the computer the nature of the operation required and leaving it to do all its calculations and Boolean operations in binary, without assistance (except for the programs in ROM).

The four arithmetic operations and the Boolean operators

Computers carry out all four arithmetic operations – addition, subtraction, multiplication and division – by converting them all into addition (possibly repeated), using Boolean operators together with the carry flag. Once these simple operations are burned into the ROM, it is possible to expand the range of operations by installing algorithms to calculate such things as square roots, logarithms and trigonometric functions (sine, cosine, etc.). Mathematicians have calculated many such algorithms using only addition, multiplication, subtraction and division.

Addition

Addition is performed by applying the Boolean operator known as the exclusive OR function (written XOR). This can perform a number of tasks, such as comparing binary numbers in size and adding them. The truth table below shows the four possible outcomes. If either A OR B is positive (but not both), the value of XOR is also positive. But if both A and B are positive (or negative), the outcome is zero. (The slogan is 'either A or B, but not both'.)

	A	B	A XOR B
(i)	0	0	0
(ii)	0	1	1
(iii)	1	0	1
(iv)	1	1	0

Of course, adding more than one bit is more complex, and will probably involve a carry flag. This is set to 1 if both bits are 1, and is otherwise 0. (Its value is given by the Boolean operator called AND: see page 213). The circuit that adds one bit with a carry flag is known as a 'full adder', and combinations of full adders are used to add whole bytes together. (The reason for the name is that the adder without carrying does only half the work needed, and so is called a half adder. Eight full adders together will add whole bytes, and are known as 8-bit adders.)

Multiplication as addition

Computers perform multiplication by repeating the operation of addition several times. For example, if you add 5 to itself six times $(5+5+5+5+5+5)$ you get 30, that is 5×6. This applies to any multiplication sum: 13×259 is the same thing as 13 added to itself 259 times. Human beings quickly become bored by this method – which is why schoolchildren learn the short cuts of multiplication tables and long multiplication, the only purpose of which is to eliminate interminable chains of addition. Computers not only cannot be bored, but are also able to work out millions of additions per second. It is simple to design circuits for multiplication: all that is needed to multiply each pair of 8-bit numbers together is a pair of 8-bit adders connected together, and a circuit to count the number of additions made.

Subtraction

The computer transforms the subtraction of one number from another by changing the number subtracted into its complement and adding. This works in binary arithmetic just as it does in decimal: for example, $255 - 128 = 255 + (-128)$. In binary, the complement of a byte is given by the Boolean operator NOT. (There are occasional problems with the generation of extra 'carry' numbers at the end, but simple refinements cure them.) For example, if the computer is asked to calculate $243 - 95$, it proceeds as follows:

        1111 0011   (243)
    $-$ 0101 1111   (95)

is the same as

            1111 0011
    $+$ NOT(0101 1111)

which is

        1111 0011
    $+$ 1010 0000

that is

    11001 0011   (403)

The extra 1 at the end (the first figure on the left) is moved round and added at the beginning (one of the refinements mentioned above):

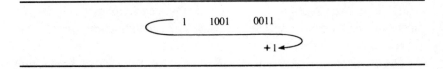

making the number we are seeking:

1001 0100   (148)

Division as subtraction

Division means subtracting one number (the divisor) as many times as possible from another (the dividend). For example, 94 can be subtracted five times from 478 leaving a remainder of 8. In other words, 478 can be divided by 94 five times, leaving 8. Using binary numbers, this can be shown in a table as follows:

478	1	1101	1110	
−94		0101	1110	
384	1	1000	0000	once
−94		0101	1110	
290	1	0010	0010	twice
−94		0101	1110	
196		1100	0100	three times
−94		0101	1110	
102		0110	0110	four times
−94		0101	1110	
8		0000	1000	five times
So	1	1101	1110	(478)
divided by		0101	1110	(94)
equals		0000	0101	(5)
remainder		0000	1000	(8)

which is $\frac{478}{94} = 5$ rem. 8

Processing the program – the CPU

The CPU (that is, central processing unit) consists of a relatively small number of registers, about five in all, which oversee the running of the program. These are the memory address register, the program counter, the accumulator or register A, the auxiliary register B and the instruction register. They are active in completing the tasks necessary to ensure the meticulous carrying-out of the instructions, checking that the correct sequence is followed, and in anticipating a malfunction due to the overloading of specific registers.

The memory address register keeps track of the program, making a note of the line numbers. It receives input from the program counter and instruction register. It receives the addresses (that is, line numbers) of words stored in the memory. The line number or program address is normally part of the instruction. The address can therefore point either to data words or to an instruction. The program counter keeps track of the address of the instruction being fetched and executed. The line number in the program counter is incremented at the same time as the instruction is executed. The accumulator and the way it functions have already been described (see page 231). Register B normally receives and supplies the second number needed in the arithmetic operation. The instruction register stores the instruction, which is fetched from the memory where it is decoded to determine the nature of the operation. The relationship between all the registers is dynamic and constantly changing, as the computer prepares for various contingencies and copes with them.

Counting by flip-flop

In mechanical calculators, which work by gears and cogs, each cog corresponds to a different number in the sequence from 0 to 9. Electronic counters count electric pulses, timed in millionths of a second by an electronic 'clock'. The process of combining (AND-ing) the changing states of the flip-flop with timed electric pulses is what enables counting to be done at all.

To be able to count we must have some way of recording and storing the number of digits or pulses, these being the discrete units of which the total is composed. If we are operating with a decimal count, we need numbers from 0 to 9. The next number counted, 10, resets the units counter to zero and also records a carry of one to the tens place. The decimal counter therefore needs at least ten elements (0 to 9) for the units place. The same is true for other places: tens, hundreds and so on. The same symbols are quite adequate to specify any size of number when combined with the place-value.

Although early computers used the decimal system, it soon became apparent that binary counting was better since there were only two digits to store, namely zeros and ones. The method of counting is by means of a flip-flop, which records a one on being 'set' and a zero on being 'reset' ('cleared', that is set to zero).

A good analogy for the way the electronic counter works is the way early humans may have counted sheep. The work is most efficiently done by two people. One person calls out the number of each sheep in the sequence between one and ten, making a tally mark on the ground for each sheep as it is herded into the field. After every 10 sheep, the second person also records a tally mark. When the gate is closed behind the last sheep, the tens and the final unit tallies are combined to make the grand total. Essentially, the electronic counter performs the same two functions. The flip-flop changes its state repeatedly, each change creating a signal that corresponds to the individual tallies. This is combined with another binary signal from the timer. Both signals pass through an AND gate. The timer generates a pulse/no-pulse sequence; the other signal consists of 'set' and 'reset' pulses which signal the alternating stable states of the flip-flop. This all happens in a few millionths of a second.

The future of number and machine computation

There are two main strands to computer research. One is the discovery of new applications of computing and the development of those applications that already exist. The other is the development of more powerful computers.

In the area of new applications, future developments are difficult to predict. This is because all the main developments made in the past appeared, as it were, out of the blue: someone thought up a new way in which computers could make life easier. Some of the main lines of current research are described below.

Simulations of cells and biological processes

In Chapter 16 I speculated about Alan Turing's second research interest, morphogenesis (that is, the origins of the shapes of bodies). For example, we might ask, 'Why is a kangaroo a different shape from a human being?' and 'What makes an octopus different in appearance from a polar bear?' In answering such questions, we must deal with two main problems. First, there is the problem of evolution (that is, the response of the organism to external forces, especially the struggle for existence). Second, there is the problem of internal forces (that is, how the development of each cell affects all the others). A computer mathematician, such as Turing, would seek to invent a calculus and a new kind of computer-based simulation of the effects that these problems have on cell formation and development. There is no doubt that computer simulations, already used in virtually every branch of knowledge and human activity, will remain a prime research and teaching area in the development of number.

Artificial intelligence

AI (the search for artificial intelligence) has become one of the most important areas of research into computer applications. At the moment, computers can perform only 'routine' tasks, in spite of the fact that the routines can be extraordinarily complicated. But as soon as an unforeseen problem or fault comes up, the best they can do is to stop and signal for human assistance. (They 'beep', following Babbage's idea of the analytical machine ringing a bell for the operator.) If an 'intelligent' computer – one that would pass the Turing test (see page 224) – were ever developed, then it could improvise a solution to such a problem just as easily, if not more so, than any human operator.

The closest approximations to artificial intelligence today are probably the learning computers. These are programmed with the rules for a game (the most successful ones play chess), and then play against each other or against human beings, learning all the time from their mistakes. Research into such machines has tremendous implications for improvements in teaching in schools, kindergartens, universities and the home.

Bio-chips

Bio-chips are one development that is almost certainly a long way in the future. They are computers whose structure incorporates living material. The idea is that this will give them an element of unpredictability which will enable them to call on intuition when necessary – in other words, to 'think' a whole problem through, in context. (This is also a major goal of artificial intelligence research.)

Ethical problems

In all AI and bio-chip research, ethical problems are a major challenge. Many people believe that any attempt to produce a computer with the mind of a human being is blasphemous in the extreme. Apart from this reaction (which affects even unsuccessful research), there are other, even more difficult problems to address. Would a computer that could pass the Turing test of appearing to be human be entitled to the same rights as a human being? Would there be movements for the emancipation of computers (and users) from the slavery of programs? At present, such problems are only ideas; the solutions are a long way off. They appear now mainly in science-fiction stories in which the computer rebels against human control and creates havoc. (The best known example is probably HAL in the book and film *2001*.) Revolt of this kind is extremely unlikely: enough controls would be put on the system by its developers as to make such conduct impossible – we hope!

The power of computers

Mathematicians have long since shown that the set of calculations that a computer can perform is strictly limited, and that the simplest computer (the Turing machine) can mimic all the functions of any more complex device. Because of this, a more powerful computer cannot do more; it can merely do the same things faster. This means that research in this field is aimed at designing computers that will run more operations per second. Computer engineers have designed circuits to run more efficiently (for example, by designing computers that are job-specific, that is, they will performs only limited operations). They have also invented smaller and smaller components (silicon chips replacing transistors replacing valves – or 'tubes' as Americans call them), and have modernised the methods of producing them.

Laser computers

In any electrical circuit, there is a flow of electrons. Although it may appear that, for example, a light begins to shine the instant you switch it on, there is actually a microscopic time lag, because electrons do not travel at the speed of light. In fact, they travel at about the speed of sound, approximately one-thousandth of the speed of light. Modern computers are almost at the point where this, rather than the size of their components or the efficiency of their layout, is the factor limiting the speed of their calculation. The idea of a light computer is to replace the silicon chip with miniature laser beams, thus raising the speed of computation one thousandfold. The technology has not yet been perfected, but it seems that, in a few years' time, it will be, and is likely to make a big impact.

Parallel processors

Another recent development to speed up computation is the parallel processor. In an ordinary computer, the processing is done serially: that is, the processor performs a sequence of operations, in turn, one at a time. A parallel processor works on a totally different principle, whereby a number of serial processors operate independently, each performing one part only of the whole sequence of operations. As they all do their work at once, this clearly speeds up the calculations. Two parallel processors halve the time of calculation; 25 make it one twenty-fifth what it was before, and so on. (This is, of course, the mass-production method that so fascinated Babbage.)

This technology is in an early stage of development. Several problems need to be overcome. For example, it is not always possible for the processors to be totally independent: for practical purposes, the results of one operation will normally affect another. This means that the processors will need to communicate in some way. This both provides a formidable challenge in itself, and leads to further problems. The biggest of these is that the most efficient way to connect the processors for one job may well be different for another. Any general-purpose parallel processor must have some way of changing the connections between its component serial processors. At present, this problem means that parallel-processing computers are extremely job-specific, designed to make only one kind of computation, but with maximum speed and efficiency.

Smart cards

'Smart cards' are a development of the credit card, itself an unforeseen spin-off of research into data storage on magnetic strips. A smart card will contain a miniature computer. It will store such information as telephone numbers, engagements, bank transactions, medical and social-security records. Already, such a card can be produced. The problem still to be overcome is the development of cards that will be cheap enough and small enough. The solutions to such problems, however, are only a matter of time.

Fractal geometry

Another main area of advance, already becoming more and more visible, is fractal geometry. Fractals are a tool that allows us to prepare visuals (that is, graphics) which represent natural objects or other phenomena. Fractals are composed from standard equations and reiterated calculations of a simple kind (simple for the computer, that is). The results are among the most beautiful and complex works of art ever produced as a result of human thought and expertise. They show that, contrary to some preconceptions, mathematics is anything but dull. (There are no dull subjects, only dull teachers.) Mandelbrot is the name to conjure within this area.

The importance of the computer

All calculating aids, from algorithms to log tables, from *quipus* to electronic spreadsheets, were constructed to speed up and simplify our handling of number. They assist us in all the aspects of human living, that is, in our practical activities as well as our research into the construction and working of the universe and ourselves, and in the ordering and manipulation of abstract ideas. These activities are possible – and indeed have been accomplished in earlier times – without sophisticated number-tools and computer hardware, but also without the modern panache. Throughout history, there has been a direct connection between the level of number-work in societies (its sophistication and speed) and the level of 'civilisation'. Of course, the existence of number expertise, and of the tools that assist it, is no guarantee of 'high' civilisation, but their absence has proved an effective barrier against it.

At a simple level, the computer (both in its 'mainframe' and 'desk-top' forms) is the latest in a long line of such calculation aids. It allows number-work and, by extension, all alpha-numeric activities to proceed at a speed and with an accuracy unimaginable in any previous generation. One computer (located in an office or bank, say) can do the work that previously took the time and energy of hundreds, sometimes thousands, of clerks. In 20 minutes, a computer can perform calculations that once required as many years (for instance, creating logarithm tables). Yet, when computers were invented they were the object of considerable suspicion, not only from laypeople but from those who should have known better, including teachers and professional mathematicians. Now that most of us routinely accept computers, we use our reliance on their speed and accuracy to let us 'spin off' into hitherto inaccessible or awesomely complex realms of number-work, and of the science that depends on it.

But computers are far more than mere calculation tools. They differ from every other previous device in a way that gives them infinite potential. They can 'understand', and manipulate, anything at all so long as it can be turned into numbers. They can be programmed to process words, charts, graphs, and images of every kind. They make decisions (for example, when to buy or sell stocks and shares). They run sequential operations of every kind, from manufacturing processes to space flights. They store records, help with medical examinations and diagnoses, and facilitate work in design, engineering and the arts. In a computer-oriented society, almost no aspect of human life is left untouched. To take a simple example, in the USA, using a computer service, you can dial from home to access all the airlines in North America for the cheapest return fare to, say, Hawaii, and book a seat.

Doing this costs less than taking a taxi downtown to a travel agency, with its severely limited facilities – and is obviously more convenient.

In short, in the last 40 years or so, computers have speeded up and facilitated human life to an unprecedented degree. In that time, the human race has developed intellectually more than in the whole of the previous millennium. Whether this means that we have reached some kind of pinnacle of civilisation, compared to our computerless ancestors, is a question as imponderable as what will happen next. Only one thing seems certain about future generations: that they will think of us now, in computer terms, in much the same way as we look back on the megalith-builders with their neolithic inch and neolithic yard.

THE NATURE OF SCIENTIFIC CHANGE

Science is the great antidote to the poison of enthusiasm and superstition.
Adam Smith, 1776

The simplest schoolboy is now familiar with truths for which Archimedes
would have sacrificed his life. *Ernest Renan, 1883*

Scientific revolutions

Every so often in the political history of human societies there are relatively
short, sharp periods of revolutionary change. Power relations within
society are challenged by some subversive movement. When the challenge
becomes substantial enough to be inconvenient, it begets a violent reaction
from the ruling group. This results in a period of escalating violence and
bloodshed. Depending on the actual power relations, the revolutionary
forces may seize power, or the forces hostile to change may eliminate them.
The relative strengths of the opposing forces, their levels of commitment in
military and political enterprise, the allies that each side can call upon, and
the support that they enlist when the fighting begins, are the factors that
pretty well decide the outcome.

In science, too, there are periods of revolution, which share some features
of political revolution. Lacking the incitements to physical violence and the
bloody struggles of social and political revolution, a scientific revolution is
more like a formal, ongoing dialogue between opposing parties. Loyalty
oaths are uncommon; there is rarely, if ever, an appeal to arms. The
discussion takes the form of reasoned argument with few threats, moderate
language being used by all disputants. The Hemholtz–Brücke physiological
pact of the 1840s ('that no forces other than the normal physical and
chemical ones would be called upon to explain the functioning of living
matter') was intended as an anti-Müller, anti-vitalist manifesto put

together by Müller's brightest students. (Müller was a leading exponent of the vitalist doctrine, that life on Earth could not have originated from chemical and physical forces alone, but must have been triggered by some external, 'vitalising' principle.) Its formality and its commitment make it an unusual document in the history of science.

Scientific revolutions are usually the result of informal alliances with an occasional 'set piece' in the form of a review, programme or manifesto designed to sharpen the awareness of the professionals in a particular area. An example that springs to mind is the debate on Darwinian evolution between Thomas Henry Huxley and Bishop Wilberforce at the 1860 meeting of the British Association in Oxford (see page 247).

The process of change

There have been several attempts to explain the origins and process of social revolution. The model is clear. There is a conflict of interest between two groups or social classes over the division of social resources (normally the surplus of wealth, opportunity, or political power). The ruling group, by virtue of its control of resources, has a monopoly of violence (legal or extra-legal), which it uses to suppress opposition. In states where there is (for whatever reason) no possibility of democratic change, there are no effective political means that can be used to thwart the establishment's monopoly and misuse of force, except perhaps passive resistance. In such circumstances, the alternative is for the oppressed to take to the streets and seize power by taking over key institutions of government: parliamentary offices, ministries, business and industrial establishments, police and military forces, the media, and so on. In a successful revolution, transfer of power normally involves little actual violence. This is because the people at large agree that the displaced power system has forfeited its right to govern. The majority will have experienced its violence and corruption, or their effects, at first hand.

Revolutions in science and mathematics, by comparison, usually begin from the perception that the existing body of knowledge is flawed. It fails to explain key problems, or has reached a point where further advances are, or seem to be, impossible. The value system on which scientific inquiry is based is one that, traditionally, favours change and openness. Scientists are expected to consider their results from the standpoint of reasoned enquiry. To be more specific, they are expected to base their conclusions on the results of applying the criteria of inductive logic to the detail of new experiments designed to test accepted explanations of phenomena.

There are, however, impediments to all ideal systems of values and procedures. Since modern science developed a career structure based on the setting up of training courses for students and on specific prerequisites for faculty appointments and promotions, it has become organised in terms of such things as job patterns, career opportunities, national societies, research grants, industrial and business links and affiliations. A hierarchy, based ostensibly on merit, has developed in each of the scientific occupations. In order of importance, its criteria are, first, age, with special reference to length of service in prestigious occupations and institutions; second, ability, as shown objectively by 'breakthroughs' in scientific understanding; third, judgements by peers of the 'authoritative' nature of the scientist's contribution. In recent years this differential has been called – a little wryly, perhaps – 'professor power'.

Another important factor opposing change is ideology. Teachers, by their identification with existing and 'established' knowledge, tend to be among the more conservative groups in the population. Radicalism in younger members of the profession is confronted by the conservatism of place-holders. The impasse created by these power relations has led to the stereotype view, first expressed by Max Planck, that science advances not because new discoveries are accepted after the arguments in favour have prevailed in rational debate, but because the order generation of scientists sooner or later retires or passes away.

A further barrier to change is the influence of religious beliefs and superstition. Striking examples from the past are the baneful influence, through the centuries, of Platonism and Pythagoras on the history of 'Western' mathematics, and the unwavering hostility of the mediaeval Christian Church towards 'free' scientific thinking of any kind. In modern times, attempts at thought control by religious leaders are more successfully resisted. Those who set themselves up as authorities, divinely appointed, to censor scientific views that they perceive to be in conflict with assertions in religious texts about nature, God and human beings, are the object of amusement rather than fear. Huxley's sharp reply to Bishop Wilberforce, in the 1860 debate on evolution, that if he had the choice between being descended from an ape or from a Bishop who prostituted his talents by debating subjects about which he knew nothing, he would choose the ape every time, seems in retrospect to breach the norms of legitimate scientific comment. But it is regarded by scientists as a landmark in the argument betwen science and 'religion'.

In our own time, one result of the success of Darwin's theory of evolution in transforming the natural, biological and virtually all other sciences is that we have become accustomed to making a further distinction: between 'normal' science and the scientific paradigm. Thomas Kuhn pointed out

	Early paradigm	Displaced by	'Normal' activities
Astronomy	Ptolemaic system	Copernican system	Star charts
Geology	'Terra firma'	Continental drift	Mapping surveys
Physics	Newton's system	Relativity	Ballistics, for example
Natural Sciences	'Creationism'	Darwinism	Description of species and varieties
Chemistry	Phlogiston theory of combustion	Oxygen theory of combustion	Analysis of compounds
Mathematics	Greek geometry	Non-Euclidean geometry	Accounting and computing

(correctly) that there are strong elements of mythology in the picture that scientists have of the history of their special subject and of the history of science in general. Contrary to popular belief, scientific research consists normally of humdrum, routine tasks: chiefly, measurements of one kind or another, or the checking and re-checking of facts that are already known with a high degree of certainty. Such tasks pre-empt activity for long periods and engage the majority of scientific stalwarts, who know, and need to know, little about the history of their special subject or its place in the scheme of things.

'Normal' science is carried out within a theoretical framework of assumptions about the nature of reality, and the limits of the methods used to study it, with in-depth understanding of a highly selected, specialised area of the science. This body of knowledge, opinion and ideology is the 'paradigm'. It is the frame of reference, or context, within which science develops.

Occasionally, quite unpredictably, while involved in some task imposed by 'normal' science, the investigator notices an anomaly, such as the failure of some expected result, or some departure from protocol. Realising that this may be breaking new ground, and having alerted his or her associates (to stake a claim, as it were), the investigator then proceeds to examine the situation further. Such work may lead nowhere, or it may cut across the accepted paradigm – in which case, after further study and a great deal of discussion, the paradigm is changed to take account of it.

At an earlier period, this revolutionary process was recognised, not in science but in philosophical discussion. The neo-Kantians referred to the paradigm 'shift' as a change of *Weltanschauung* (that is, in the general apprehension of the nature of reality). It was, they said, 'the transvaluation of all values'.

Some paradigmatic shifts in science are listed in the table on page 248. Some are the result of abandoning religious explanations of the creation of the universe; others are the result of more practical, specialised work.

Spheres of influence

Just as forms of government, national alliances and rivalries, legal systems and regional boundaries change over the centuries, so do the patterns of knowledge, their development and their organisation. Science is also affected by external factors, which are associated with paradigm changes, in the same way as it is by internal factors. In time, the discrepancies between new observations and outmoded explanations become more and more apparent and unacceptable.

Since about 1789 (the beginning of the French Revolution), the idea has been prominent all over the world that the proper unit of political concern and historical analysis is the nation state. In ordinary, 'lay' terms, we habitually think and write and talk about, for example, the contemporary French, or German, contribution to chemistry, the British, or Chinese, influence on mathematics, the American, or Soviet, achievements in space travel.

It is easy to accept that political relations between different nation states can be explained by the theory of 'spheres of influence'. Strong states overwhelm weaker ones; vulnerable nations form alliances against a common enemy. One state may exert an influence because it is regarded as a leader in dress fashions, life-style or cultural (including scientific) innovation. The concept of the 'client state' is well known to economists and historians. At all periods of history, some nations dominate and others are subservient.

Scientists regard such national specifics as descriptive terms only. They are reluctant to infer that there are any real national differences in science. If such differences seem genuinely to exist, they are historical accidents and there is nothing systematic about them. The record shows many examples of the same discovery being made (and reported as original) many times throughout history, or of simultaneous discoveries, in unconnected parts of the world, of the same law or principle. In mathematics, for example, Ruffino's method of solving equations of high degree was known to the ancient Chinese. Omar Khayyam, and many before him, knew the arithmetic triangle, which was later credited to Pascal. Three people believed that they discovered non-Euclidean geometry (see page 86) – but priority is given to Euler. Leibniz did not 'discover' the differential calculus (Newton did) or invent the binary system (the Chinese did). Furthermore, his basic philosophy (Monadology) was 'borrowed' from (and attributed by him to) Lady Anne Conway.

In short, so far as the history of science is concerned, the post-1789 idea of nation state is too narrow a concept. As the record shows, the views and allegiances of scientists, their informal organisation in 'schools' and more formal organisation in national societies are affected by a far larger phenomenon, what can be called the *ekumenai* or 'spheres of influence'. The word was coined in ancient Greece to imply 'all people anywhere who think like us', and was later used – as its English derivative 'ecumenical' suggests – by Christians to describe the totality of Christians, of all kinds and in all countries: a loose geographical, but tight conceptual unity. The word applies to science in a similar sense. In the history of number, for example, as this book has shown, the spheres of influence at one period or another have been those of Babylonian, Greek, Chinese, Indian, Arab and English mathematics.

There have been perhaps a dozen or so periods of such dominance in the intellectual history of the human race. One group or another acts as pacemaker in a particular area of human expertise. Sometimes the cause is historical accident. For example, the Arab political empire of the 8th to 15th centuries influenced many nations – Syrians, Turks, Egyptians, Tunisians, Persians, Iraqis. It led to a kind of pan-Arab identity, and to a remarkable cultural and intellectual unity. The satellite nations, as they had become, adopted the Arabic language, the religion of the Koran and the scientific culture of their overlords.

So far as mathematics is concerned, the *ekumenai* can be listed as follows:

1. Neolithic, pre-literate European and North American
2. Sumerian, Babylonian, Akkadian, Chaldean, Phoenician
3. Egyptian
4. Greek/Roman – first *Pax Romana*
5. Chinese, Japanese, Korean
6. Arab, Indian, Syrian – first Scientific Revolution
7. Roman Catholic, Mediaeval European – Second *Pax Romana*
8. Renaissance, Christian, European (especially Italian)
9. Western European (especially British, French and German) – second Scientific Revolution
10. Old European science and mathematics; world science

It is clear from this list that objective intellectual inquiry must have been the last thing in the minds of many of the peoples who made, or led, the great advances. Some of these nations were military adventurists; others were entrepreneurs; others had vast wealth to manage, from local resources or shipped from abroad. In almost every case, the primary urge, indeed the only urge, was to develop number skills as efficient practical tools. Those thinkers who made number their main study often did so in the service of convoluted and inept notions (nothing to do with science) of how the universe had been formed and of the supernatural conditions necessary for its continuance. Taking the high moral ground, we have good reason to condemn the conquerors, imperialists, freebooters and others whose activities simultaneously girdled and enslaved the world. Whether the benefits of their culture – which include a more scientific and realistic view of the universe by a minority – were passed on to the societies that they colonised is arguable. Equally significant is the fact that, while modern life is shaped by developing science, this still remains a closed book to most of us.

The future of mathematics

There is no doubt that, thanks to the invention of the electronic computer, and in view of developments discussed earlier, we will soon be unable to talk about number in the old-fashioned way, in the language to which we have been accustomed. Already, among the new generation of students the brightest (not necessarily the most vocal) have a new vision and are informed by a different sense of values. Following the bloodless revolution, symbolised by the computer as the centre-piece of the 'information society', a new order is struggling to be born. The old order changes, but too slowly for the present generation of scientists.

The key process is liberation from the domination of age-old, hallowed traditions and authorities. This is especially true of that most conservative of domains, mathematics: in its assumptions, its theoretical basis, and the method of teaching. A new spirit invests this area of study and practice; a process of change is in train due to the take-over of most intellectual and business functions by the computer. We can see the shape of the future in the most prosperous countries – Japan, the USA and Germany. (Whether or not we like what we see is a different question entirely, and beyond the scope of this book.)

BIBLIOGRAPHY

This bibliography is a selection of the books I have found most useful in building up a picture of numbers in early and later civilisations. In forming my views on the history of mathematics, I decided at an early stage to abandon the traditional evaluations as being based mostly on ideology and national (and professional) chauvinism rather than on scholarship. In this matter, I found much assistance and support in the two journals, *Isis* and *Osiris*, founded by George Sarton. But nothing can replace reading the primary sources. Most of the originals are available in translation; the Greeks have been especially favoured in the 'Dover Reprints' series but there are numerous reprints of other original works. Three sources were especially useful, and should be mentioned above all others: (i) Menninger's account of *Number Words* – published in English translation by MIT, Cambridge, Mass.; (ii) Joseph Needham's *Science and Civilisation in China*, especially volumes 1 and 2, and (iii) the 18-volume multinational reference work edited by C.C. Gillespie, *The Dictionary of Scientific Biography*. Other books, on specific topics and mostly written for the general reader, are listed below.

Aaboe, Asger, *Episodes from the Early History of Mathematics* (Random House, 1964)

Ball, W.W. Rouse, *A Short Account of the History of Mathematics* (Dover, 1960)

Bell, E.T., *Men of Mathematics* (McGraw Hill, 1945)

Boole, George, *Collected Logical Works* (Open Court, 1952)

Boorstin, D.J., *The Discoverers* (Random House, 1983)

Bowring, J., *The Decimal System*

Boyer, Carl B., *A History of Mathematics* (John Wiley, 1968)

Brumbaugh, Robert, *Plato's Mathematical Imagination* (Indiana UP, 1954)

Bunt, Lucas, Jones, Philip and Bedient, Jack, *Historical Roots of Elementary Mathematics* (Prentice Hall, 1976)

Burnham, R., *Burnham's Celestial Handbook* (Dover, 1978)

Cajori, Florian, *A History of Mathematical Notations* (Open Court, Chicago, 1928)

Cardano, G., *The Book on Games of Chance* (Holt, Rinehart and Winston, 1961)
The Book of My Life (Dover, 1962)

Ceram, C.W., *Gods, Graves and Scholars* (Knopf, 1952)

Chace, A.B. (ed), *The Rhind Mathematical Papyrus* (Math. Assoc. of America, 2 vols. 1927, 1929)

Childe, V.G., *Man Makes Himself* (Thinker's Library, 1938)

Clagett, Marshall, *Greek Science in Antiquity* (Abelard-Schuman, 1956) da Vinci, Leonardo, *Philosophical Diary* (Philosophical Library, 1959)

Dantzig, Tobias, *Number: The Language of Science* (Macmillan, 1954)

Datta, B. and Singh, A.N., *History of Hindu Mathematics* (Asia Publishing House, 1962)

David, F.N., *Games, Gods and Gambling: the Origins and History of Probability* (Griffin, 1962)

Dickson, Leonard E., *History of the Theory of Numbers* (Chelsea Reprint, 1951)

Dijksterhuis, E.J., *The Mechanisation of the World Picture* (OUP, 1964)

Farrington, Benjamin, *Greek Science: its Meaning for Us* (Penguin, 2 vols, 1944, 1949)

Fibonacci, Leonardo, *Liber Abaci* (British Library, 1857)

Galilei, Galileo, *Dialogue on the Great World Systems* (1632; trans. University of Chicago Press, 1953)

Gerhardt, Helen, *The Phoenicians* (Morrow, New York)

Gillings, Richard J., *Mathematics in the Time of the Pharaohs* (MIT Press, Cambridge, Mass., 1972)

Glaser, A., *History of Binary Arithmetic*

Goldstine, H.H., *The Computer from Pascal to von Neumann* (Princeton UP, 1972)

Gomperz, Theodor, *Greek Thinkers* (John Murray, 4 vols, 1901–12)

Goodwater, Leanna, *Women in Antiquity: An Annotated Bibliography* (Scarecrow Press, Mechosen, NJ, 1975)

Guthrie, W.K.C., *A History of Greek Philosophy* (CUP, 6 vols, 1962–81)

Hall, Alfred, R., *The Scientific Revolution 1500–1800* (Longmans Green, 1954)

Hamilton, Edith, *The Greek Way to Western Civilisation* (Norton, New York, 1964)

Hardy, G.H., *A Mathematician's Apology* (CUP, 1967)

Heath, Thomas L., *A History of Greek Mathematics* (OUP, 2 vols, 1920)
  A Manual of Greek Mathematics (Dover, 1963)
  Mathematics in Aristotle (OUP, 1949)
  Diophantus of Alexandria (Dover, 1964)
  The Works of Archimedes (Dover, 3 vols, 1953)

Hill, G.F., *Arabic Numerals in Europe* (New York, 1915)

Hodges, A., *Alan Turing, The Enigma* (Burnett Books, 1983)

Holton, Gerald, *Thematic Origins of Scientific Thought* (Harvard UP, 1973)

Howson, Geoffrey, *A History of Mathematics Education in England* (CUP, 1982)

Infeld, L., *Whom the Gods Love: The Story of Évariste Galois*

Irwin, Constance, *Fair Gods and Stone Faces* (St Martin's Press, 1963)

Jaeger, Werner, *Paideia* (trans. OUP, 3 vols, 1939–45)

Kahn, D., *The Code Breakers* (Macmillan, 1967)

Karpinski, Louis G., *The History of Arithmetic* (Rand-McNally, 1925)

Klein, Felix, *Elementary Mathematics from an Advanced Standpoint* (Macmillan, 2 vols, 1932, 1939)

Kline, Morris, *Mathematics: A Cultural Approach* (Addison-Wesley, Reading, Mass.)

Knott, G.G., *Napier Tercentenary Memorial Volume* (Longmans Green, 1915)

Krupp, E.C., *Echoes of the Ancient Skies: The Astronomy of Lost Civilisations* (1983)

Libbrecht, Ulrich, *Chinese Mathematics in the 13th Century* (MIT, Cambridge, Mass, 1973)

Luce, J.V., *The End of Atlantis: New Light on an Old Legend* (Thames and Hudson, 1969)

Masiarz, Edward and Greenwood, Thomas, *The Birth of Mathematics in the Age of Plato* (Am. Research Council, 1964)

Menninger, Karl, *Number Words* (MIT, Cambridge, Mass, 1969)

Meschkowski, H., *The Evolution of Mathematical Thought* (Holden-Day, 1965)
  The Ways of Thought of Great Mathematicians (Holden-Day, 1964)

Mikami, Y., *Japanese Mathematics* (1914)

Moore, Doris Langley, *Ada, Countess of Lovelace* (John Murray, 1977)

Needham, Joseph, *Science and Civilisation in China* (CUP, vol. 3, 1970)

Neugebauer, O., *The Exact Sciences in Antiquity* (Princeton, 1952)

Newton, Isaac, *Mathematical Principles of Natural Philosophy* (University of California Press, 1946)
  'Arithmetica Universalis' trans, in *Mathematical Works*, ed D.T. Whiteside, vol. 2

O'Leary, de Lacy, *How Greek Science Passed to the Arabs* (Routledge and Kegan Paul, 1949)

Ore, O., Cardano, *The Gambling Scholar* (Princeton UP, 1953)

Pannekoek, Antonie, *History of Astronomy* (Allen and Unwin, 1961)

Pullan, I.M., *History of the Abacus* (Hutchinson, 1968)

Peet, T.E., *The Rhind Papyrus* (Ohio, 1929)

Pappus, *La collection mathématique* (Blanchard, 2 vols)

Plato, Dialogues, trans Jowett (5 vols, OUP)

Proclus, *A Commentary on the First Book of Euclid's Elements* (Princeton UP, 1970)

Randall, J.H., *The Making of the Modern Mind* (Houghton Mifflin, 1940)

Rorbrough, Lynn, *Primitive Games*

Sarton, George A., *History of Science* (Harvard, 3 vols, 1952–9)

Scott, Joseph F., *A History of Mathematics* (Taylor and Francis, 1958)

Shore, Bruce, *The New Electronics*

Smith, David E., *A Source Book in Mathematics* (Dover, 2 vols, 1959)
  A History of Mathematics (Dover, 2 vols, 1958)
  and Karspinski L.C., *The Hindu-Arabic Numerals* (Ginn and Company, Boston, 1911)

Smith, Preserved, *A History of Modern Culture* (Holt, Rinehart and Winston, 2 vols, 1940)

Sommerville, D.M.Y., *The Elements of Non-Euclidean Geometry* (Dover, 1958)

Srinivasiengar, C.N., *The History of Ancient Indian Mathematics* (1967)

Struik, D.J., *A Source Book in Mathematics 1200–1800* (Harvard UP, 1969)
  A Concise History of Mathematics (Dover, 1958)

Taylor, L.G.R., *Mathematical Practitioners of Tudor and Stuart England*
  Mathematical Practitioners of Hanoverian England (1714–1840) (CUP, 1966)

Thom, Alexander, *Megaliths* (OUP, 1971)
  Megalithic Lunar Observatories (OUP, 1971)

Thorndike, Lynn, *The History of Magic and Experimental Science* (vols 1 and 2, Columbia UP, 1923)

Todhunter, Isaac, *A History of the Mathematical Theory of Probability* (Chelsea Reprint, 1965)

van der Waerden, B.L., *Science Awakening* (Noordhoff, 1954)

von Neumann, John, *Collected Works* (Pergamon, 6 vols, 1961–3)

Wallace, W.A., *Causality and Scientific Explanation* (University of Michigan Press, 2 vols, 1974)

Wedberg, Anders, *Plato's Philosophy of Mathematics* (Almqvist and Wiksell, 1955)

Weizenbaum, Joseph, *Computer Power and Human Reason* (W.H. Freeman, San Francisco, 1976)

White, Andrew Dickson, *A History of the Warfare of Science and Theology in Christendom* (Dover, 2 vols, 1960)

Whitehead, Alfred N., *Science and the Modern World* (CUP, 1927)

Wolf, Abraham, *A History of Science, Technology and Philosophy in the 16th and 17th Centuries* (Allen and Unwin, 1962)

Yushkevitch, Adolf, *A History of Mathematics in the Middle Ages* (CUP, 1985)

Zaslavsky, Claudia, *Africa Counts* (Prindle, Weber and Schmidt, Boston, 1973)

INDEX